Biomass Production and Efficient Utilization for Energy Generation

About the Authors

Prof. N. S. Rathore is Vice Chancellor, Maharana Pratap University of Agriculture and Technology, Udaipur. He did M.Tech with specialization in Energy Studies from I.I.T., Delhi and Ph.D. from Rajasthan Agricultural University, Bikaner. He has served the ICAR as Deputy Director General (Engg.), Deputy Director General (Education). He is also founder Vice Chancellor of Sri Karan Narendra Agriculture University, Jobner- Jaipur, Rajasthan. He has contributed more than 200 technical papers and 36 books on various aspects of Renewable Energy, Environment and Agricultural Engineering. He has undertaken 50 research projects and organized 20 International trainings, Seminars/winter, Workshops and Conferences and 12 ISTE sponsored Summer School. There are 100 reports, proceedings and popular articles to his credit. He was honored by the Indian Society of Agricultural Engineers and other government and private organization.

Dr. N.L. Panwar is an Assistant Professor of Renewable Energy Engineering in the Faculty of Engineering at MPUAT, Udaipur, Rajasthan (India). He has been awarded PhD from the Centre for Energy Studies, Indian Institute of Technology Delhi, India. His areas of interest cover energy and energy analyses of thermal systems and energy efficiency & management. He has contributed more than 100 papers in international journals and several books on renewable energy aspect. He has been a recipient of '*Prakritik Urja Puraskar*' from Ministry of New and Renewable Energy, Government of India for his outstanding book on Alternative Energy Resources. He has also been awarded by "Shrimati Vijay-Usha Sodha Research Award" from Indian Institute of Technology Delhi in 2014, and Rajasthan Energy Conservation Award 2018 from Government of Rajasthan.

Biomass Production and Efficient Utilization for Energy Generation

N.S. Rathore
Vice Chancellor
Maharana Pratap University of Agriculture and Technology
Udaipur (Rajasthan) India

Formerly
Deputy Director General (Education)
Deputy Director General (Agricultural Engineering)
Indian Council of Agricultural Research
New Delhi

N.L. Panwar
Department of Renewable Energy Engineering
Maharana Pratap University of Agriculture and Technology
Udaipur (Rajasthan) India

CRC Press is an imprint of the
Taylor & Francis Group, an **informa** business

NEW INDIA PUBLISHING AGENCY
New Delhi – 110 034

First published 2022
by CRC Press
2 Park Square, Milton Park, Abingdon, Oxon, OX14 4RN

and by CRC Press
6000 Broken Sound Parkway NW, Suite 300, Boca Raton, FL 33487-2742

CRC Press is an imprint of Informa UK Limited

British Library Cataloguing-in-Publication Data
A catalogue record for this book is available from the British Library

Library of Congress Cataloging-in-Publication Data
A catalog record has been requested

ISBN: 978-1-032-15811-2 (hbk)
ISBN: 978-1-003-24576-6 (ebk)

DOI: 10.1201/9781003245766

Preface

Energy crises and environmental pollution are the two major challenges before mankind in present context. The growing use of fossil fuels such as coal and oil has resulted in increasing the greenhouse gas emission. To resolve the energy crisis and reduce greenhouse gas emission, the dependency on fossil fuels shall have to be minimized by replacing them with renewable fuels. Biomass is the storage of solar energy in chemical form in plant and animal materials. It is one of commonly used, precious and versatile resources on the earth. It is rich source for feed, fodder, fiber, fuel, and fertilizer. Biomass has been used for energy purposes ever since man discovered fire.

The content of book includes all major aspects of biomass production and efficient utilization for energy generation. Most of the information presented in this book reflects a basis to acquire the understanding of the proper utilization of biomass for heat and power generation. In this book, design criteria, present state of art of technology and future perspective of clean energy are illustrated through graphs, figures, tables, flowcharts. equation etc. to make subject more clear and useful.

In this book an attempt has been made to bring out the state of art of technology on production and efficient utilization of clean and green fuel from biomass. The book has been divided into ten chapters. Chapter 1 deals with overview of biomass, energy plantation, proximate and ultimate analysis of biomass. Chapter 2 pertains the application of biomass for domestic application which include improved cookstoves and biomass-based water heating system. Chapter 3 brings out the biogas generation from biomass. Biochar production and its applications highlighted in Chapter 4. Chapter 5 deals with pyrolysis process where biomass conversion into liquid fuel is included. Design and development of torrefaction unit presented in Chapter 6. Design of biomass gasifier and its application for thermal energy generation along with practical exposure are presented in Chapter 7. Chapter 8 pertains the production of biodiesel and case study of biodiesel produced using caster seed oil in actual use. Chapter 9 deals with bioethanol production. Densification of loose biomass is presented in chapter 10. It is hoped that this book will be useful as text book and a reference book for students pursuing course in energy studies.

Udaipur **Authors**

Preface

Contents

1

Biomass- An Overview

1.1 General

The term "biomass" generally refers to renewable organic matter generated by plants through photosynthesis. Materials having organic combustible matter is also referred under biomass.Biomass is an important fuel source in our overall energy scenario. Biomass is produced through chemical storage of solar energy in plants and other organic matter as a result of photosynthesis. During this process conversion of solar energy in sugar and starch, which are energy rich compounds takes place. The chemical reaction of photosynthesis can be written as:

$$6CO_2 + 6H_2O + sunlight \rightarrow C_6H_{12}O_6 + 6O_2 + 636\,kcal$$

It indicates that the storage of 636 kcal is associated with the transfer of 72 gm carbon into organic matter. Biomass can be directly utilized as fuel or can be converted through different routes into useful forms of fuel. In fact, biomass is a source of five useful agents, which start with 'F' like food, fodder, fuel, fiber and fertilizer. Further, biomass has many advantages like.

(1) It is widely available

(2) Its technology for production and conversion is well understood.

(3) It is suitable for small or large applications

(4) Its production and utilisation requires only low light intensity and low temperature (5 to 35°C)

(5) It incorporates advantage of storage and transportation

(6) Comparatively, it is associated with low or negligible pollution.

1.2 Biomass Classification

Biomass includes plantation that produces energy crops, natural vegetable growth and organic wastes and residues. The biomass classification is illustrated in Fig. 1.1. It can be grouped as:

(1) **Agricultural & Forestry Residues:** Silviculture Crops.

(2) **Herbaceous Crops:** Weeds and Napier grass.

(3) **Aquatic and marine biomass:** Algae, water hyacinth, aquatic weeds, plants, sea grass beds, kelp and coral reef etc.

(4) **Wastes:** Municipal solid waste, municipal sewage sludge, animal waste and industrial waste etc.

India produces about 320 million tonnes of agricultural residues every year. Similarly, 273 million cattle population produces on an average about 433 million tonnes of dung annually. Fuel wood is another major source of biomass in India. The fuel wood consumption in India is estimated to be about 227 million tonnes per year. Some of biomass sources are given below:

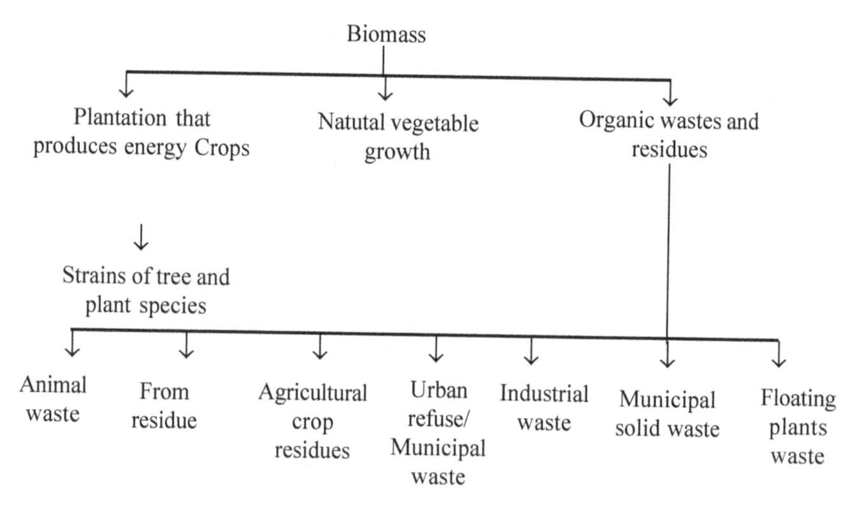

Fig. 1.1: Biomass and Its Classification

1.2.1 Energy Plantation

This term refers to an area that is used to grow biomass for energy purposes. The idea behind energy plantation programme is to grow selected strains of tree and plant species on a short rotation system on waste or arable land. The sources of energy plantation depend on the availability of land & water and careful management of the plants. As far as suitability of land for energy plantation is concerned following criterion is used

(1) It should have minimum of 60-cm annual precipitation

(2) Any arable land having slope equal to or less than 30% is suitable for energy plantation.

The economics of energy plantation depends on the cost of planting and availability of market for fuel. Whereas these two factors are location specific and these varies from place to place. Further productivity of this programme depends on the microclimate of the locality, the choice of the species, the planting spacing, the inputs available and the age of harvest. There are many suitable species for energy plantation, few of them are :

(1) *Acacia nilotica* (Babul),

(2) *Acacia auriculiformis* (Bengali Babul),

(3) *Dalbargia sissoo* (Shisham),

(4) *Eucalyptus comaldulusis* (Eucalyptus),

(5) *Leucaena lencocephala* (Subabul),

(6) *Prosopis chilensis* (Perdesi Babul),

(7) *Prosopis Juliflora* (Vilayati Babul),

(8) *Tamarix articulate* (Jungle jalebi),

(9) *Tamarindusindica* (Imali),

(10) *Albizzalebbek* (siris).

1.3 Biomass Characteristics

Biomass can be characterized for its utility and different energy usage. The various characteristics of biomass falls under following categories

(a) Proximate analysis

(b) Ultimate analysis

(c) Ash deformation and fusion temperature

(d) Calorific value

(e) Rate of devolatilization

(f) Bulk density.

These properties varies from the species to species, their moisture content as well as method employed for fuel preparation.

1.3.1 Proximate Analysis

It is the method for measuring various properties of biomass such as moisture content, fixed carbon, volatile matter and ash. The moisture content is one of the important properties of biomass, over which its heating value depends. The moisture content is determined by drying the weighed amount of sample in an open crucible kept at 110ºC in an oven for one hour. Always biomass sample is first grind to form fine powder, then this powdered sample is kept for determination of proximate analysis. In fact, for most of the biomass, the consistency in weight is obtained within one hour at 110°C or so. If required, the period of heating may be increased till the consistency of weight is obtained. The difference between the initial and final weights is taken as moisture content of the fuel.

The experimentation of moisture content determination is extended for measurement of ash content of biomass. The sample, so obtained after determination of moisture content, is then gradually heated to 750°C in a muffle furnace and is kept for two hour or were till constant weight is recorded. The weight of the residue represents the ash content of the biomass.

Volatile matter is determined by keeping the dried sample in a closed crucible at 600°C for six minutes and then at 900°C for another six minutes. The difference in the weight due to the loss of volatiles in taken as the total volatile matter present in the biomass. The fixed carbon content is found by applying the mass balance for the biomass sample. The carbon content determined through this method is not the actual carbon content present in biomass but only the non volatile part of carbon content, as some of the carbon present in biomass also escapes along with the volatile.

Measurement

(a) Weight of empty silica crucible A

(b) Weight of crucible and sample B

(c) Weight of crucible + sample after drying at 110°C C

(d) Weight of crucible + ash after ignition D

The sample calculation for measurement of proximate analysis is as follows:

(a) Moisture content % $= \frac{B-C}{B-A} \times 100$

(b) Total solid % $= \frac{C-D}{B-A} \times 100$

(c) Total volatile solids % $= \frac{C-D}{B-A} \times 100$

Moisture content on wet basis

$$MC_{wb} = \frac{Wet\ weight\ matter - Dry\ weight\ matter}{Wet\ weight\ matter} \times 100\%$$

Moisture content on dry basis

$$MC_{db} = \frac{Wet\ weight\ matter}{Dry\ weight\ matter} \times 100\%$$

Example 1.1:*Prosopis Juliflora* is having about 35% moisture content on wet basis. The ideal moisture content to gasified is about 12%. Calculate the amount of water to be removed from per tons of *prosopis juliflora.*

$$MC_{wb} = \frac{1000 - Dry\ weight\ matter}{1000} \times 100$$

Dry weight of *prosopis juliflora* = 650 kg

Weight after drying at 12% moisture on wet basis is

$$12 = \frac{Wet\ weight\ matter - 650}{Wet\ weight\ matter} \times 100$$

Wet wight is 738.6 kg

Hence amount of water to be removed from prosopis juliflora to make it suitable for gasification is:

= weight at 35% moisture content – weight at 12% maisture content

= 1000 – 738.6

= 261.4 kg of water

1.3.2 Ultimate Analysis

The ultimate analysis gives carbon, hydrogen, oxygen, nitrogen, and sulphur contents of the fuel. C-H-O analyser determines the carbon and hydrogen contents by standard method. Further, knowing the ash content, oxygen is determined by difference. However, the samples must be dried prior to analysis. Nitrogen and sulphur are normally negligible.

C-H-O analyser is essentially consists of (i) an electric furnace (ii) a sample column and (iii) absorbent column. The dry matter is powdered and weighed (w_1) before putting it in the sample column. The absorbent column is filled with a weighed quantity (w_2) of calcium hydroxide. Subsequently the furnace is started and oxygen from a separate oxygen cylinder is supplied to the sample column at a pressure of 4 PSL. A temperature of more than 1400°C is maintained for about 20 minutes. Then the furnace is switched off and the fused sample is taken out and weighed (w_3). The calcium hydroxide from the absorbent column is also taken out and reweighed (w_4). From these observations the carbon content of the sample can be determined using the following relationship. The difference (w_4-w_2) will give carbon dioxide formed.

$$\text{Carbon in absorbent } (w_5) = \frac{w_4 - w_2}{w_3} \times 12$$

$$\% \text{ Carbon in the sample} = \frac{w_5}{w_1} \times 100$$

1.3.3 Ash Deformation and Fusion Temperature

There is a standard test for measurement of fusibility of coal and coke. It is based on ASTM D 1857. In this method, first of all biomass is dried and grounded and finally placed in the muffle furnace at 750°C in the presence of air till constant weight is obtained. The residual ash so obtained is then finely ground. Then this ash is converted in plastic mass by adding a solution containing 10% dextrin, 0.1% salicylic acid and 89.9% H_2O by weight. This plastic mass is moulded to a cone shape by pressing it into a suitable mould. Then, these canes are taken out and allowed to dry. Those dried canes placed on a refractory base are then inserted in a high temperature furnace to about 800°C. After about 15 minutes interval the temperature of the sample is raised at an increment of 50°C. During each interval the shape of the cane is observed. The temperature at which initial rounding off or bending of the apex of the cone is observed is known as ash deformation temperature. If this temperature is further increased, the same sample would fuse into a hemispherical lump. The temperature during this phenomenon is known as "ash fusion temperature".

1.3.4 Heating value

The heating value of the oven dried biomass samples is determined by Bomb Calorimeter method. The heat evolved when unit mass of fuel is burnt is known as higher calorific value (HHV). The Bomb Calorimeter method is used for determining higher heating value of biomass. Whereas, the lower heating value (LHV) is calculated by subtracting the heat liberated during condensation of the water vapour formed due to combustion of hydrogen content of the fuel. The hydrogen content is known by the ultimate analysis. The lower heating value is also known as net heating value. Normally net heating value is utilized in combustion and gasification process. The lower heating value can be estimated using following formulae:

$$MC_{wb} = \frac{1000 - Dry\,weight\,matter}{1000} \times 100$$

where H, M, and hg are hydrogen percentage, moisture percentage, and latent heat of vaporization respectively.

1.3.5 Bulk Density

It depends upon the moisture content, shape and size of the biomass. As this property normally varies depending upon the fuel preparation method employed, therefore it should be determined in site for specific applications.

1.3.6 Rate of Devolatilisation

The rate of devolatilisation of biomass can be determined by a standard Thermogravimetric Analyser (TGA). The TGA has a furnace with linear heating rate as 4 °C per minute. Also, in this unit, a desired gas flow over the sample can also maintained. Generally, N_2 gas at a flow rate of 300 cm^3/min is used to create the oxygen free inert atmosphere to avoid combustion of biomass. The unit is continuously records the loss in the weight of the sample because of thermal decomposition as a function of time and temperature. From these datas, the percentage conversion is calculated at different temperature. The percentage weight loss per degree rise in temperature can be used for defining exact degree of pyrolysis process.

Biomass is somewhat complex organic matter. However, for the purpose of energy stored it can be regarded as a mixture of carbon, hydrogen and oxygen.

The main constituents of any biomass material are:

(i) Lignin

(ii) Hemi cellulose/Xylan

(iii) Cellulose

(iv) Mineral matter

The percentage of these components varies from species to species. The composition to typical biomass species is given in Table 1.1.

Table 1.1: Composition of Biomass

Species	Cellulose	Hemi Cellulose	Lignin	Ash
Soft wood	41	24	27.8	0.4
Hard wood	39	35	19.5	0.3
Rice Straw	30.2	24.5	11.9	16.1
Bagasse	33.6	29	18.5	2.3

Biomass composition in terms of proximate analysis, ultimate analysis and other properties for typical fuels are given in Table 1.2 to 1.4.

Table 1.2: Proximate Analysis

	Coal	Sawdust	Groundnut shell	Rice Husk
Moisture (%)	2 - 4	8 – 10	11 - 13	8 – 10
Volatile matter (%)	40 -45	70 – 75	65 - 72	50 - 55
Fixed carbon (%)	50 - 60	15 -20	20 - 22	13 -15
Ash (%)	4 -6	2 -4	2 - 5	20 - 23

Table 1.3: Ultimate (Elemental) Analysis

Species	C (%)	H (%)	N (%)	O (%)	Ash (%)
Wood	44-52	5-7	0.5-0.9	40-48	1-3
Rice Husk	33 - 37	5-6	0.5 – 0.6	36 - 38	20 -23
Bagasse	46 - 49	4 - 6	0.1 – 0.2	40 -45	3 - 4
Groundnut Shell	45 - 53	4- 6	05 – 1.5	38 -45	2 – 5

On an average biomass consists of 40-50% carbon, 4-7% Hydrogen and 30-45% Oxygen on moisture and ash free basis. Biomass contains negligible amounts of nitrogen and sulphur.

Chemical formula on the basis of elemental analysis of wood:

Let us C =44 %; H = 5 %; N = 0.7%; O = 42%

Assuming 100 g wood sample

$$H = \frac{5}{1} = 5.00$$

$$H = \frac{5}{1} = 5.00$$

$$N = \frac{07}{14} = 0.05$$

$$O = \frac{42}{16} = 2.62$$

Chemical formula for wood is:

$C_{3.66}H_5N_{0.05}O_{2.62}$

Table 1.4: Typical Comparative Properties

	Coal	Wood	Rice Husk
Bulk density (kg/m^3)	700 - 900	0.2 0.35	300-400
Ash fusion temperature (K)	900 - 1200	1200 - 1400	1400-1650
Calorific Value (MJ/kg)	17-23	18-21	13-17

1.4 Biomass Production Technique

The biomass production should be view with number of points starting from preparation of soil to planting seedlings and upto harvesting of the biomass. In fact, these all operations are combined for biomass production to achieve multiple benefits. Thus, a careful planning is required for biomass production, which consists of integration of different techniques and improved methods. The general sequence for biomass production, which consists of integration of different techniques and improved methods. The general sequence for biomass production is illustrated in Fig. 1.2. It includes step starting from site survey, nursery techniques, transplanting techniques and maintenance of the plantation.

1.4.1 Site Survey

As far as site survey for biomass production is concerned, it should watch with the species of tree and shrubs best suited to the area. Many information are needed for proper site survey, such as climate, soil, topography, vegetation, biotic factors, water table levels, availability of supplementary water sources and distance from nursery etc.

1.4.2 Planting Site Selection

It is based on the information of the site survey. The best site is one, which on plantation will lead to the establishment of a successful plantation.

However, the choice of the planting site is limited to lands, which are not suited for agriculture or livestock production. The boundaries of the planning site should be marked with boundary posts after the area has been chosen. When there is a danger of trespassing and damage by grazing animals a boundary fence should be established. Of course, fencing is costly and therefore it should only be built when other means of protection are not effective. Once the forest plantation is well established with sufficiently tall trees, the fences can be removed and reused at another planting site. When roads and other passageways traverse the planting site, they also should be contained with fences.

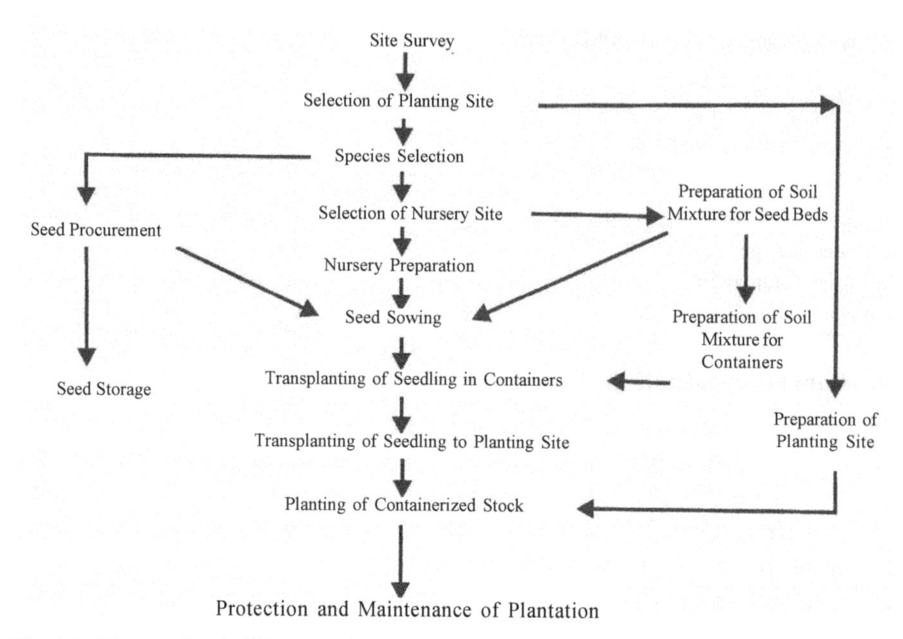

Fig. 1.2: Biomass Production Techniques

1.4.3 Species Selection

Once the selection of site plantation is over, then following points should be consider in the selection of the tree and species.

1. A species selected for a particular site, should suit to the site and should remain healthy throughout the anticipated rotation and should exhibit acceptable growth and yield.
2. For a successful planting, performance data can have to be extrapolated from one locality to another. Results from a locality where a tree or shrub species is growing (either naturally or as an exotic) strictly apply only to that locality.
3. When reliable information shows a close similarity between the site to be planted and that on which the species already is successful, this particular site is recognized for largescale planting.
4. The selection of tree and shrub species through the use of analogous climates is important as a first step in selection. This step is amplified by an evaluation of localized factors (for example soil, slope and biotic factors).
5. The ability to match closely a planting site and a natural habitat may not preclude the need for species trials, since climatological or ecological metering may not reveal the adaptability of a species.

1.4.4 Preparation of the Planting Site

The preparation of the planting site is more important in the biomass establishment programme.The site preparation is meant for removing competing vegetation from the site, creating conditions that will enable the soil to catch and absorb as much rainfall as possible and providing good conditions for the planting, including a sufficient volume of routable soil. The site preparation also includes the creation of conditions where the dangers from fire and pests are minimized. The site preparation is directed toward giving the seedlings a good start with rapid early growth. It reduces surface runoff for increasing the moisture in the soil and at the same time it helps in elimination of hard pans of the soil.

1.4.5 Preparation of the Soil Mixture

The soil mixture for the containers as well as for seedbeds should have following characteristics.

(a) It should be relatively light and cohesive

(b) It should have a good water retention capacity

(c) It should possess a high organic and mineral content

(d) It should contain adequate nutrients, which usually are supplied in the form of artificial fertilizers.

(e) The quantity of soil needed in a containerized nursery operation is directly related to the size of the containers used. For example, to fill 1,00,000 small containers, 200 cubic meters of soil are needed, where as 442 cubic meters of soil are needed to fill 1,00,000 of the largest containers (16 times more).

(f) The Soil used in nursery work should be sufficiently acidic in nature. For most tree and shrub species, the pH should not be higher than 6.0.

Damping-off is a common and serious disease in many forest nurseries. It can occur either in seedbeds or in containers after transplanting. This disease is caused various fungi. A watery - appearing construction of the stem at the ground line generally is visible evidence of the disease. Damping off is favored by high humidity, damp soil surface, heavy soil, cloudy weather, an excess of shade, a devise stand of seedlings and alkaline conditions.

One of the best preventive measures for this disease is to maintain dry soil surface through cultivation, to reduce the sowing density and to thin the seedling to create better aeration at the ground line.

1.4.6 Sowing of Seed

Sowing of seed should match to the ideal sowing time. In order to determine the ideal sowing time, one counts backwards from the beginning of the planting season to identify the number of months required to raise high quality planting stock. The time to raise high quality planting stack depends on number of factors like; type of tree or shrub species, climatic conditions and nursery conditions.

It is essential for each planting project to prepare its own sowing schedules for the locally important tree and shrub species. With large amount of seeds, not all should be sown at the same time. It is better to spread out the sowing dates with one or two week intervals.

A common method of raising seedlings has been to sow the seed in seedbeds or seedling trays, and then to transplant into containers as soon as the plants are sufficiently large to handle. The seedbeds can be either broadcast sown or sown in lines. Small seed should be mixed with some kind of inert fine material to facilitate even distribution.

1.4.7 Method of Sowing

Seedling trays or boxes commonly used for the production of seedlings, which subsequently are transplanted into containers. The trays can be easily moved to the beds for the transplanting operation. A good tray can be made from kerosene or diesel tins by cutting them lengthwise into halves. The bottom of the tray should be perforated to allow drainage or irrigation by absorption. In order to assist drainage, the bottom of the tray can be filled with a 2 cm layer of gravel or charcoal.

Once the seeds are sown in seedlings or seedling trays, they should be covered with sand, which should be pressed down gently to establish a good contact between seeds and soil. Depth of sowing depends on the size of the seed. The normal depth of sowing is from 1 to 3 times the diameter of the seed, but in many instances, it can be desirable to sow slightly deeper to avoid the washing of seeds from the soil by irrigation or heavy rains.

1.4.8 Transplanting of Seedling into Containers

A plant that is grown either in seedbed or seedling trays, in which it was originally sown and than it has to lifted for final replacement, is technically known as a “seedlings”. When a seedling is lifted and replanted in the nursery in another bed or container, it is, therefore, termed a “transplant”. The transplanting of seedlings, also referred to as “pricking” which is done primarily to induce of better development of the root system by increasing the number of fine absorbing root lets.

Transplanting should be done before the seedling has acquired a large, heavy root system, but after it has developed a strong stem. Normally, this stage occurs after the complete unfolding of the cotyledons and during the unfolding of the first leaves. Transplanting is a delicate and time-consuming operation and requires precise organization and sufficient labour. Therefore, there is a tendency to sow directly into the containers when possible, which has the advantage of saving time and labor, and reducing the losses and retardation in plant growth caused by transplanting.

1.4.9 Transport of Seedlings to the Planting Site

Generally, plants are damaged during transport to the planting site. Therefore, adequate care must be taken to avoid mishandling of plants during loading and unloading from vehicles. Even, plants require protection during transportation, as the airflow can cause breakage of stems or drying. Therefore, it is also important that the containers as packed tightly, so that they cannot move. When possible, plants be transported in the planting season on cool, cloudy or even rainy day to prevent desiccation during transport. Following are few suggestions, which should be kept in mind during transportation of seedlings to the planting site.

(a) Always prefer the packing of container raised plants for transportation.

(b) Put packed container raised plants in trays the load into vehicles.

(c) The tins, which have been used for seedling trays, can be used for transporting container plants.

(d) Shipping schedule should be planned to avoid delays and to allow proper disposition of the plants immediately upon arrival.

(e) Normally, plants should arrive one day ahead of planting.

(f) Where shape and watering facilities are available, supplies can be brought in advance of several days.

(g) As soon as the plants arrive at the planting site, they must be watered and if necessary, heeled in a cool, moist, shaded place until they are needed for planting.

1.4.10 Maintenance of the Plantations

Once a plantation has been established, than it is necessary to protect the plantation against damaging. A variety of cultural treatments also can be required to meet the desired purpose of the plantation Protection of plantation against weather effect, fire damage, pests and insects, wild animals is essential. The cultural treatments include weeding, thinning, pruning and maintaining desired spacing between trees.

1.5 Harvesting of Biomass

Biomass harvesting is decided when trees and shrubs attain the "optimum size" for the wood production. From a biological point of view, trees and shrubs should not be cut until they have at least grown to the maximum size required for product utilisation. In general, the average annual growth of trees and shrubs increases slowly during the initial year of establishment, reaches a maximum and then falls more gradually in subsequent years. Hence, trees and shrubs usually should not be allowed to grow beyond the point of maximum average annual growth, which is the age of maximum productivity. As far as harvesting of biomass is concerned following factors be considered.

1. Biological factors -Age of maximum productivity, average growth rate.
2. Pathological factors-Growth in terms of mortality and the amount of defect in living trees.
3. Entomological factors-Forest composition, age, structure and vigour.
4. Silvicultural factors -Seed production, characteristics, methods of obtaining regeneration, competition from less desirable tree species and maintenance of desirable soil conditions.
5. Local harvesting techniques.
6. Available man power.
7. Existing market outlets.

Biomass harvesting should be well planned and organized in order to make the best use of the raw material while keeping labour input and production cost low and minimising damage to the environment. A variety of different harvesting systems can be applied. The factors on which harvesting systems depend are as follows

1. Species of wood
2. Size and assortment (fuel wood, poles or logs)
3. Type of forest (man made or natural)
4. Type of cut (thinning or clear cut)
5. Kind of regeneration (artificial, coppice, natural)
6. Terrain (flat, steep, swampy)

7. Accessibility (roads, waterways) and
8. Means of transport (manual, animal, motorized)

In all these cases, good planning and organization of work depends on following factors

1. The assessment of the volume to be harvested.
2. The determination of the assortments to be produced.
3. The determination of wood storing places, skidding lines and felling direction.
4. Clearly-instructed and skilled supervision and workers.
5. Availability of the necessary hand tools and maintenance and other equipment.
6. Clear separation of working areas for individual work teams and different operations (filling and transport).

1.5.1 Harvesting Methods

These are several different harvesting methods that allow the plant to re generate through sprouting. These are as follows

(A) Coppicing

It is one of the most widely used harvesting methods in which the tree is cut at the base, usually between 15 - 75 cm above the ground level. New shoot developed from the stamp or root. These shoots are some times referred to as sucker or sprouts. Management of sprouts should be carried out according to use. For fuel wood the number of sprouts to container to grow, should depend on desired sizes of fuelwood. If many sprouts will be allowed to grow for a long period, the weights of the sprouts will become heavy and the sprouts may tear away from the main trunk. Several rotations of coppicing are usually possible with many species. The length of the rotation period depends on the required tree products from the plantation. It is suitable method for production of fuelwood. Most eucalyptus species and many species for leguminous family, most of naturally accessing shrubs can be harvested by coppicing (Fig. 1.3a).

(B) Pollarding

It is the harvesting systems, in which the branches including the top of the tree are cut, at a height of about 2 meter above the ground and the main trunk is allowed standing. The new shoots sprout or emerged from the main stem to develop a new crown. This results into continuous increase in the diameter of main stem although not in height. Finally when the tree losses its sprouting vigour, the main stem is also cut for use as large diameter poles. An advantage of this method over coppicing is that the new shoots are high enough off the ground that they are out of reach of most grazing animals. The neem tree (*Azadirachata Indica*) is usually harvested in this manner. The branches may be used for poles and fuel wood (Fig. 1.3 b).

(C) Lopping

In this method most of the branches of the tree are cut. The fresh foliage starts sprouting from the bottom to top of the denuded stem in spite of severe defoliation, surprisingly quickly. The crown also regrows and after a few years, the tree is lopped again. The lopped trunk continues to grow and increase in height, unless this is deliberately prevented by pruning it at the top (Fig. 1.3c).

(D) Pruning

It is very common harvesting method. It involves the cutting of smaller branches and stems. The clipped materials constitute a major source of biomass for fuel and other purposes, such as fodder mulching between tree sows. It is also often required for the maintenance for fruits and forages trees, alley cropping and lives fences. The process of pruning also increases the business of trees and shrubs for bio fencing. Root pruning at a distance of 2 - 3 meter from the hole is effective to reduce border tree competition for water and nutrient with the crops (Fig. 1.3 d).

(E) Thinning

It is a traditional forestry practice and in fuelwood plantation, it can also be of importance. The primary objectives of the thinning are to enhance diametric growth of some specific trees through early removal of poor and diseased tree to improve the plantation by reducing the competition for light and nutrients. Depending on initial plant density, initial thinning can be used for fuelwood or pole production. If maximum biomass production is the main objectives, of the plantation, regardless of quality, thinning may not be needed.

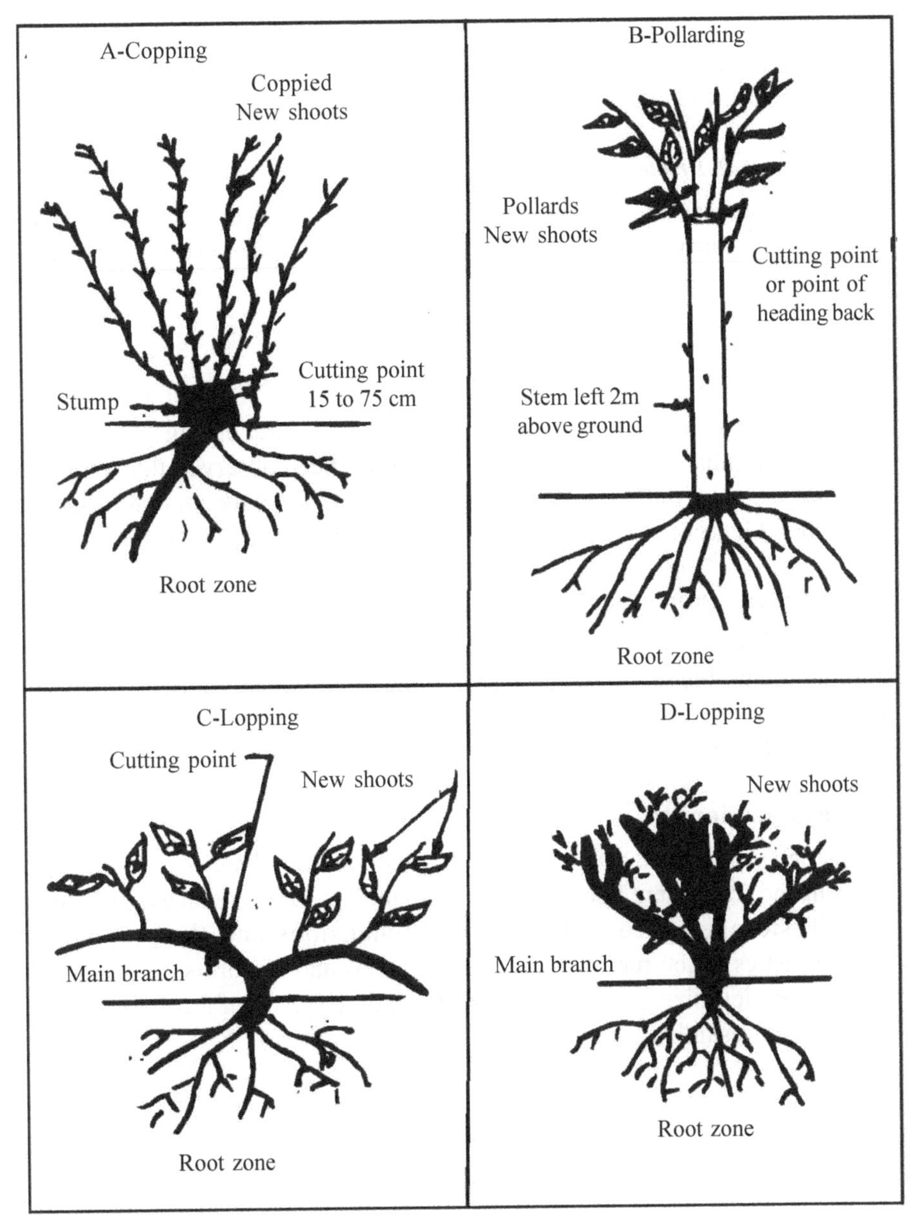

Fig. 1.3: Different Harvesting Methods

1.6. Biomass Processing for Fuel Use

Biomass can be processed into a number of ways for fuel use. The biomass can be used as an energy source material for heating, mechanical work or for electricity generation. It can be combusted directly to produce heat or it can be converted into a secondary fuel or into an energy carrier, such as steam, compressed air, electricity etc. Secondary conversion is through physical, chemical or biological processes or a combination of these. General biomass processing routes is shown in Fig.1.4.

1.6.1 Physical Processes

Physical processes involves many unit operations such as drying (change in the moisture constant), size reduction (change in surface area to volume ratio) and densification (change in density).

(a) Drying

It is method of removing moisture content of biomass. The heat generation of biomass is dependent on the moisture content, therefore, drying of biomass is essential for enhancing its calorific value. It has been observed that about 9% of energy value of biomass are lost in reducing the moisture content from 30% to 9%, but if it is not reduced, the decrease in calorific value is about 26 percent.

(b) Size Reduction

The size reduction of biomass is done to convert it to a convenient transportable, storable and usable form. These processes include tree cutting (removing stems and branches from tree), log cutting to domestic size or convenient size for use and log cutting to small billet form for use.

(c) Densification

As the name indicates, this process is to increase bulk density of biomass for efficient and convenient transportation and handling. It also reduces requirement of bulk storage space. The physical dimensions and the combustion characteristics of the fuel become homogeneous and uniforms because of the required particle size, porosity and density and as a result of it efficient energy conversion is established.

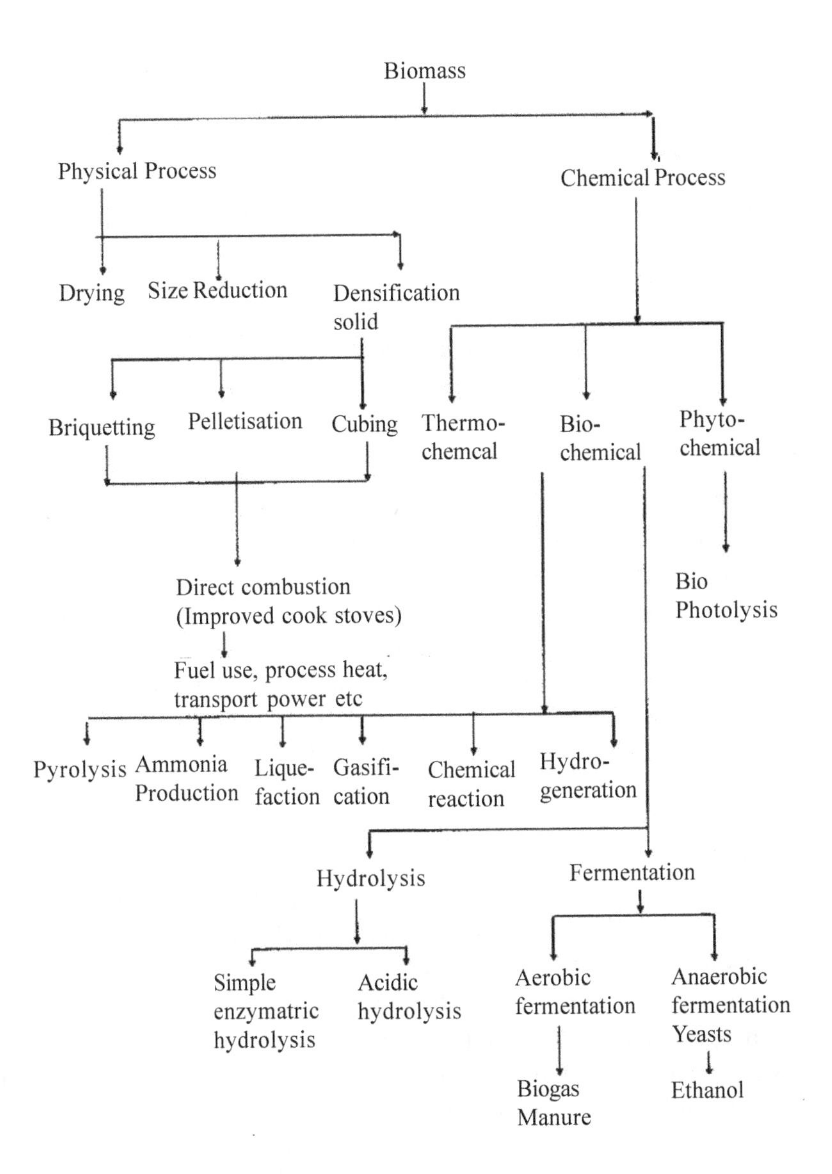

Fig. 1.4: Different Biomass Processing Processes

Densification processes are basically of three types

(i) Binderless densifcation technique - Where no external binders are added for compacting the materials. However, it is done at very high temperature (200°C) and at high pressure (1500 kg/cm^2).

(ii) Binder densification technique - Where pressure requirement is less through same types of binding materials such as tar, molasses, sodium bentonite, resins or wax.

(iii) Pyrolysis - densification technique - In this technique, firstly the biomass is carbonized and the charcoal obtained is powdered and then compacted in required shape.

Depending upon final shape of product obtained through densification process, the process is terms as briquetting, pelletization and cubing.

(d) Biomass Briquetting

Bulk of the energy in the rural areas for cooking and heating purposes is derived from firewood, agro and forest residues and to some extent from cow dung. Use of firewood leads to deforestation and inefficient combustion of cow dung, agro and forest residues leads to pollution problems. In addition, materials like pine needles are major causes of forest fires.

To solve these problems and to save wood, the agro and forest residues can be upgraded to convenient and smokeless briquetted fuels. These materials can be converted into small sized briquettes with holes (beehive briquettes) through partial pyrolysis process. These briquettes can burn with a smokeless and clean flame, ideal for domestic and small scale heating purposes. The agro and forest residues can also be directly briquetted by compaction at high pressures. These briquettes are suitable for large scale and industrial applications. The agro and forest residues in briquetted fuel form has added advantage of easy transportation to the place of work.

1.7 Animal Waste

The animal waste is also an organic material, which have combustible property. The amount of wastes an animal can produce depends on the size of the animal, feed and the condition of confinement. The animal waste is a rich source of fuel. The dung cakes so prepared with animal wastes can be used for meeting cooking energy requirement in rural and semi urban areas. Further, recycling of animal waste anaerobically can produce a combustible gas known as biogas. Thus, animal waste is a essential ingredient for fuel production through cakes and anaerobic digestion. The anaerobic digestion is a process through which not only combustible gas is obtained, but also enriched manure is produced which can help in replacement of chemical fertilizer. The production of dung from an animal is generally assumed to be directly proportional to the live weight of the animal. Average dropping from each animal species is depended on the physical characteristics of animal. It is given in Table 1.5

Table 1.5: Average Dropping by Animal

S.No.	Animal	Normal average dropping per animal per day (kg)
1.	Cow	10.0
2.	Buffalo	15.0
3.	Bullock	12.5
4.	Camel	6.0
5.	Horse	10.0
6.	Sheep	1.0
7.	Pig	2.25
8.	Chicken (2 kg weight approx)	0.18

There are a wide range of animal wastes that can be used as sources of biomass energy. The most common sources are manures from pigs, chickens and cattle (in feed lots) because these animals are reared in confined areas generating a large amount of waste in a small area. In the past this waste has been recovered and sold as a fertilizer or simply spread onto agricultural land, but the introduction of tighter environmental controls on odour and water pollution means that some form of waste management is now required, which provides further incentives for waste-to-energy conversion. In countryside the women made dung cake and it use as a fuel in house for cooking purpose the typical Indian women carrying dung cake is illustrate in Fig 1.5

Fig.1.5: Dung Cakes in India

1.8 Agricultural Residues

Large quantities of crop residues (waste matter) are produced annually worldwide, and are vastly under-utilised. A common agricultural residue is the rice husk, which makes up 25% of rice by mass. Other plant residues include sugar cane fibre (known as bagasse), coconut husks and shells, palm oil fibre, groundnut shells, and cereal straw.The annual production of crop residues in India is more than 500 million tons (Mt). Among these residues, surplus crop residues are estimated to be in the range of 84 to141 Mt/yr, in which cereals and fibre crops comprise approximately 58% and 23% of the total crop residues[1].

The annual gross potential of agricultural biomass is determined by using residue-to product ratio (RPR). In order to determine the amount of biomass produced, it is necessary to know the RPR. The potential of the entire biomass in each district of the state has been cumulated on the basis of the model[2] as presented in equation (1).

$$(CR)_t = (RPR)_t \times (PrC)_t \qquad (1)$$

where $(CR)_i$ is the amount of agricultural biomass of i_{th} crop in ton, $(RPR)_i$ the RPR of the ith crop on dry mass basis and $(PrC)_i$ the amount of crop production in ton. Main product with crop residue ratio of different crops are listed in Table 1.6

Table 1.6: Crop Residues Ratio of Different Crops

S.No	Crop	Main product	By-product	Main product to crop residue ratio (CRR)
1.	Rice	Rice	Straw Husk bran & broken	1:1.5 2.2:1
2.	Wheat	Grain	Straw	1:1.2
3.	Sorghum	Grain	Stalk	1:3
4.	Pearl millet	Grain	Stalk	1:4
5.	Maize	Grain	Stalk & cob	1:4
6.	Other cereals	Grain	Straw	1:1.5
7.	Bengal gram	Grain	Straw	1:1.3
8.	Pigeon pea	Grain	Stalk	1:4
9.	Lentil	Grain	Straw	1:2
10.	Other pulses	Grain	Straw	1:1.5
11.	Ground-nut	Kernel	Shell	1:1.1
12.	Rapeseed & mustard	Grain	Stalk	1:2
13.	Soybean	Grain	Straw	1:1
14.	Sunflower	Grain	Stalks and flower	1:2
15.	Other oilseed	Grain	Straw	1:1
16.	Cotton	Cotton	Stalk	3t/ha
17.	Jute & Mesta	Fiber	Stalk	3t/ha
18.	Sugarcane	Sugarcane stalks	Trash and green tops bagasse	10% of stalk
19.	Potato	Tuber	Haul	1t/ha
20.	Onion	Bulb	Straw	1t/ha
21.	Coconut	Kernel	Coconut shells Coconut husk	3t/ha1t/ha
22.	Cashew nut	Kernel	Husk	1:1

1.9 Generation of biofuels

First generation biofuels are produced directly from food crops. The biofuel derived from the starch, sugar, animal fats, and vegetable oil are fall under this category. It is important to note that the structure of the biofuel itself does not change between generations, but rather the source from which the fuel is derived changes. Corn, wheat, and sugar cane are the most commonly used first generation biofuel feedstock.

Second-generation biofuels, also known as advanced biofuels, are fuels that can be manufactured from various types of non-food biomass. Biomass in this context means plant materials and animal waste used especially as a source of fuel.

The term **third generation biofuel** refers to biofuel derived from algae.

The fourth-generation biofuels are made using non-arable land. However, unlike third-generation biofuels, they do not require the destruction of biomass. This class of biofuels includes electrofuels and photobiological solar fuels. Some of these fuels are carbon-neutral.

Example 1.2: Estimate the amount of crop residue generated from 15 ha of rice filed per year. The crop residue ration for rice straw is to be taken 1:1.5. if the yield is about 5 tons per ha per year.

Solution: The waste generated during rice harvesting

Total production of rice = Field size (ha) $\times$ yield (tons per ha per year)

Total production of rice = 15×5

= 75 tons per year

Waste generated

= total yield (tons per year) $\times$ CRR

= 75×1.5

= 112.5 tons per year

References

Regional Project DPR on Crop Residue Management 15 January 2018.pdf at www.moef.gov.in (assessed on December 22, 2018)

Singh J, Panesar BS, Sharma SK. Spatial availability of agricultural residue in Punjab for energy. Journal of Agricultural Engineering Today 2003; 27(3–4): 71–85.

2

Biomass Energy for Domestic Applications

2.1 Introduction

Biomass store solar energy in the chemical form and it is most precious and versatile resources on earth. Biomass, unlike fossil fuels, is a renewable energy resource that is available where the climatic conditions are favorable for plant growth and production[1]. The term biomass is used for all organic materials which are combustible in nature, mainly plant and animal origin present in land and aquatic environments. Biomass is considered carbon neutral, because the amount of carbon it can release is equivalent to the amount it absorbed during its life time. There is no net increase of carbon to the environment in the long term when combusting the lignocellulosic materials. Therefore, we can say that biomass is a renewable source of energy and can play vital role in responding to concerns over the protection of the environment and the security of energy supply[2,3].

Biomass energy accounts for about 15% of the world's primary energy consumption and about 38% of the primary energy consumption in developing countries[4]. Furthermore, biomass often accounts for more than 90% of the total rural energy supplies in developing countries[5]. It is the main source to meet out domestic energy in developing countries[6]. The efficient use of fuel wood is much more eco-friendly than the use of efficient and conventional fuels like kerosene and liquefy petroleum gas (LPG). LPG emits 15 times more CO_2 per kg than wood, and kerosene emits nearly 10 times as much. CO_2 is the prominent candidate for global. It is commonly assumed that biomass fuel cycles based on renewable in nature and greenhouse-gas (GHG) neutral because the combusted carbon in the form of CO_2 is soon taken up by re-growing vegetation[7].

The combustion process in traditional cooking stove is non-ideal and favoring incomplete combustion. Incomplete and inefficient combustion by traditional cookstoves, produce significant quantities of products of incomplete combustion

(PIC) importantly fine and ultra fine particles which have more global warming potential (GWP)[8] than CO_2.

Cooking stoves using wood fuel are the most common combustion devices in the world and are used by over 2.4 million people. Traditional cooking stoves are inefficient[9,10] and are linked to 1.6 million deaths per year from indoor pollution according to World Health orOanization (WHO). There is urgent need to speed up the dissemination of cleaner, more efficient and better ventilated stoves technology. It is the improved use of biomass in households, institutions and industries which leads to reduced fuel consumption, faster processing, improved product quality and products with better shelf life[11]. Other benefits that accrue from increased use of improved biomass stove include the alleviation of the burden placed on women and children in fuel collection, freeing up more time for women to engage in other activities, especially income generating activities. Reduced fuel collection times can also translate to increased time for education of rural children especially the girl-child[12]. The provision of more efficient stoves can reduce respiratory health problems associated with smoke emission from biofuel stoves[12]. Efforts were made to design and develop mud based improved cookstoves, which permitted higher thermal efficiency and safe removal of smoke. However, mud based improved cookstoves regularly required cleaning and polishing with mud and dung mixture. As a result, their dimensions may change and the hole designed for taking required quantity of air for proper combustion, may suck less air, hence the thermal efficiency may disturb. It was observed that average life of such mud based improved cookstoves is also less than one year on account of its material of construction and frequent cleaning and polishing. There are numbers of biomass cookstove designs are available and can be classified on the basis of their construction material, number of pot, type of fuel, etc. The classification of biomass cookstoves is depicted in Fig. 2.1.

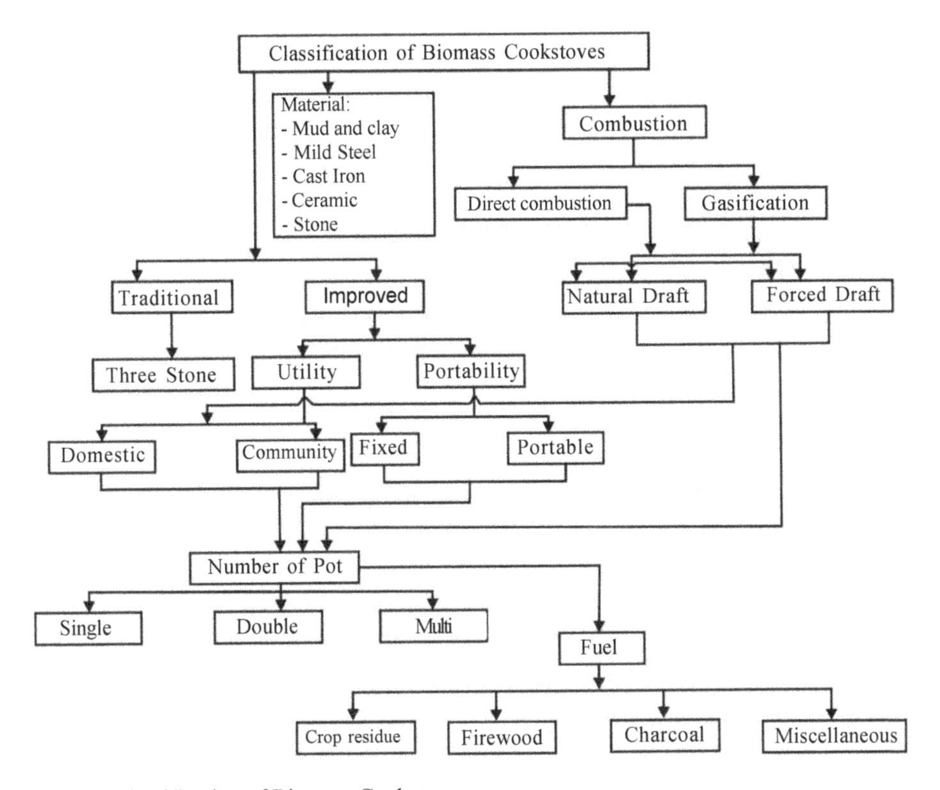

Fig. 2.1: Classification of Biomass Cookstoves

Cooking is one of the important fundamental need for human existence. The type of food and the cooking practices, reflect the cultural traditions of the civilizations inhabited this globe since time immemorial. As back as history reveals, majority of communities all over the world used wood and biomass as fuel for their cooking requirements. Department of Renewable Energy Engineering, MPUAT, Udaipur, has designed & develop single and double pot durable improved cookstoves for average sized family which have at least 5 years of useful life.

2.2 Single Pot Improved Cookstove

This cookstove is suitable for firewood, agro waste and dung cakes. Bigger sized wood can also be used in this cookstove at the fuel burning rate of 1.00 kg h^{-1}. The passage between the firebox chamber and the inlet of the chimney pipe is horizontal, which flows smoky gases outside the kitchen. A mixture of cement and sand is used to fix up the bricks and to enhance the life of the structure. The centre also developed a metallic mould for this cookstove for speedy and accurate construction in the field. The overall size of this improved cookstove is 60 x 45x 24 cm as shown in Fig. 2.2 and it technical specification presented in Table 2.1. This model is suitable for those who prefer one cooking at a time.

Table 2.1: Technical Specification of Single Pot Improved Biomass Cookstove

Parts No.	Parts Name	Material
1.	Main block	Brick, Cement, and Sand
2.	Chimney tunnel	Cement
3.	Cut portion of chimney pipe	Cement
4.	Chimney 75 mm diameter and 3 m length	Cement
5.	Pot hole	—
6.	Fire box	—
7.	Cowel	—

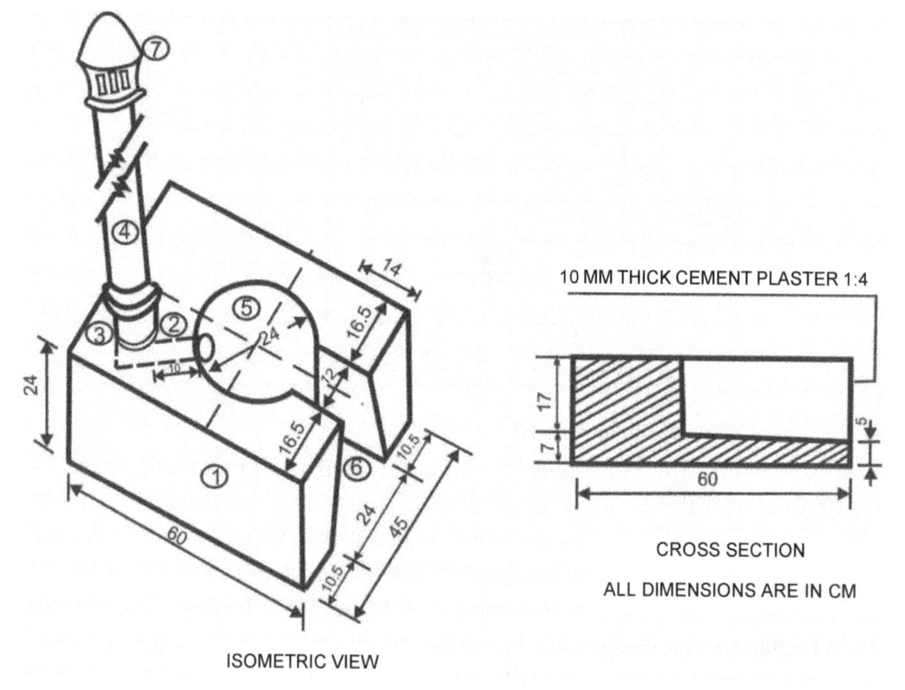

Fig. 2.2: Schematic of Single Pot Improved Biomass Cookstove

2.3 Double Pot Improved Cookstove

This model having two pot holding places is specially designed to meet the cooking requirement of the rural family. This stove is sufficient to cook the meal of a family of 6-12 persons and is of permanent nature. The special feature of this improved biomass cookstove is bigger size of fire box, which is suitable for big size firewood and light agro-waste type of fuel. The diameter of the 1st pot is 24 cm, which is suitable for households for baking of wheat flour made bread named as "Chapatti". The diameter of 2nd pot is 20 cm and suitable for cooking vegetables and boiling of milk etc. The burning of wood is very efficient without back fire. The overall size of this improved cookstove is 85 x 40 x 25 cm as shown in Fig. 2.3 and it technical specification is presented in Table 2.2.

Table 2.2: Technical Specification of Double Pot Improved Biomass Cookstove

Parts No.	Parts Name	Material
1.	Main block	Brick , Cement, and Sand
2.	First tunnel 50 mm diameter	Cement
3.	Chimney tunnel	Cement
4.	Cut portion of chimney pipe	Cement
5.	Chimney 75 mm diameter and 3 m length	Cement
6.	Cowel	Cement
7.	Place for first pot	—
8.	Fire box	—
9.	Place for second pot	—

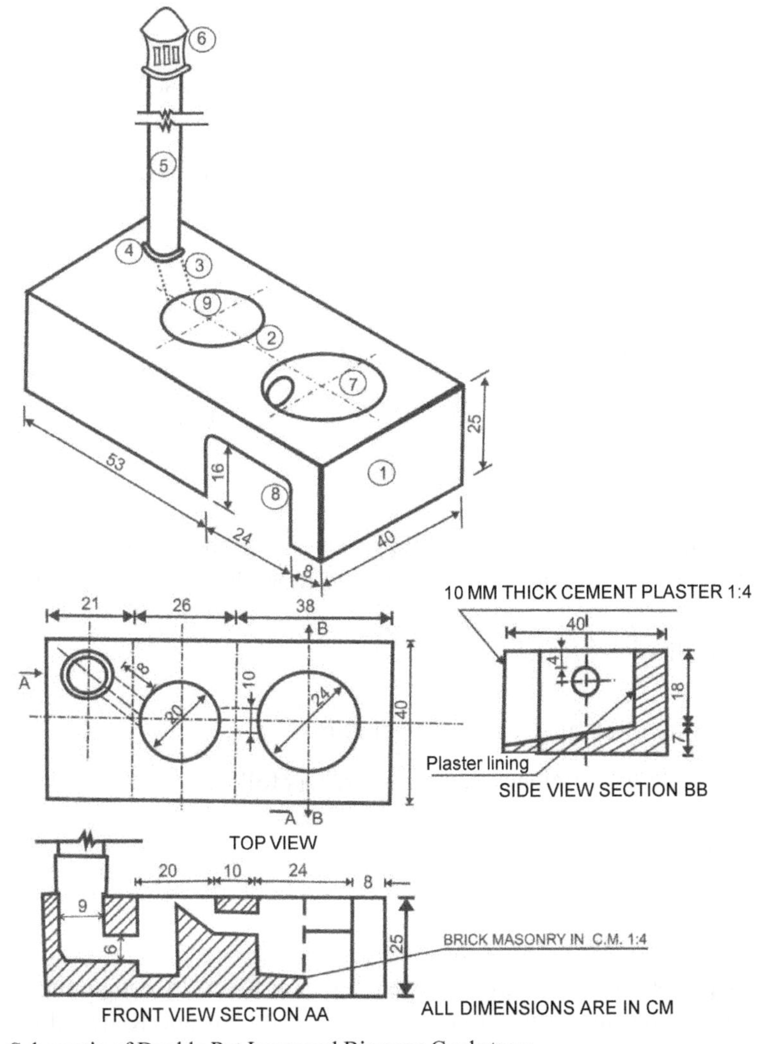

Fig. 2.3: Schematic of Double Pot Improved Biomass Cookstove

2.4 Exergy Analysis of Cookstove

Biomass is the indirect source of solar energy and it exergy contents can be calculated by using their higher heating or lower heating values. Both higher and lower heating values is the function of weight fraction (wt%, dry) of chemical composition of biomass such as Carbon (C), Hydrogen (H), Oxygen (O) and Nitrogen (N), etc. The higher heating value (HHV) of biomass can be calculated by using correlation proposed by Bilgen[13]

$$HHV = [33.5(C) + 142.3(H) - 15.4\,(O) - 14.5\,(N)]10^{-2}$$

Similarly, lower heating value can be computed by using correlation proposed by Finet[14].

$$LHV = HHV\,(1 - H_2O_i) - 2440\,(H_2O_i + 9H_i)$$

Where H_2O_i and H_i state the moisture content of biomass and the weight of hydrogen in biomass.

Szargut and Styrylska[15] correlation can be used for estimating the exergy of biomass.

$$Ex_{biomass} = \beta LHV_{biomass}$$

Where β is the quality factor and can be calculated as follows:

$$\beta = \frac{1.0412 + 0.2160\left(\frac{H}{C}\right) - 0.2499\left(\frac{O}{C}\right)\left[1 + 0.7884\left(\frac{H}{C}\right)\right] + 0.0450\left(\frac{N}{C}\right)}{1 - 0.3035\left(\frac{O}{C}\right)}$$

Exergy output of improved cookstove is depends of the heat utilized to boil the water and amount of water evaporated and it can be written as follows:

$$Ex_{out} = H_T\left(1 - \frac{T_{am}}{T_{fw}}\right)$$

Where T_{fw} state for final water temperature

Therefore, exergy efficiency can be written as follows:

$$\psi = \frac{Exergy\,output}{Exergy\,input} = \frac{Ex_{out}}{Ex_{biomass}}$$

2.5 Testing Protocol

Thermal performance of improved cookstove was assess by conducting standard water boiling test as per guideline of Ministry of New and Renewable Energy

(MNRE) formerly known as Ministry of Non-Conventional Energy Sources (MNES) which are as follows:

1. Fill the vessel with known quantity of water so as to occupy two third volume of vessel (W_1)lit.
2. Record the initial temperature (T_1 °C) of the water
3. Start the fire by ignite small piece of chopped fuel wood. As soon as the fuel wood catches fire in the pot hole, place the vessels containing water on the seats properly. Note the starting time (t_1) of the test
4. Allow the combustion of fuel in such a way that flame becomes steady.
5. Note the temperature rise of water at regular intervals.
6. When water starts boiling, note down its temperature (T_2) and the time at this stage (t_2) and remove lid.
7. Keep constant fuel bed rate until duration of the test is over. Note down timing (t_3) and final temp. (T_3).
8. Extinguish the fire in the heart immediately at the end of the test duration. Take out coal and with.
9. Measure quantity of residual water; calculate quantity of water evaporated by subtracting quantity of residual water from the initial quantity of water taken in vessel.
10. The test should be continued till two consecutive readings become constant.
11. The test should be repeated after allowing sufficient time for cooling of cookstove.

Proximate analysis of fuel was carried out before the test by using the method suggested by ASTM(1983). Fifteen trails were taken in different conditions to carry out the thermal efficiency test of the both cookstoves. Calorific value of fuel wood was calculated by Digital bomb calorimeter (Advance Research Instrument Company). Physical and thermal properties of fuel wood were presented in Table-2.3.

Table 2.3: Physical and Thermal Properties of Fuel Wood

Characteristic		Biomass (Desi Babul wood) *(Acacia nilotica)*
Size (mm)	:	25-40
Length (mm)	:	35-75
Bulk density (kg m^{-3})	:	350
Moisture content (% wb)	:	5.6
Volatile matter (% db)	:	82.52
Ash content (% db)	:	1.05
Fixed carbon (% db)	:	16.43
Calorific value (kJ kg^{-1})	:	15491 kJ/kg

Observations

Calorific Value of fuel wood : CV kJ/kg

Fuel consumption : FC kg

Quantity of water

1. First Pan : W_1
2. Second Pan : W_2

Water evaporated

1. First Pan : w_1
2. Second Pan : w_2

Rise in Temperature

1. First Pan : T_2-T_1
2. Second Pan : T_3-T_1

Heat gain by water $H_1 = W_1 C_p (T_2 - T_1) + W_2 C_p (T_3 - T_1)$

Heat Utilized to evaporate the water

$$H_2 = (w_1 + w_2) \times Latent\ heat\ of\ water\ (2260 \frac{kJ}{kg})$$

Total Heat Absorbed by water $(H_T) = H_1 + H_2$

Heat produced by fuel = $F \times CV$

$$\text{Thermal efficiency } \eta_{th} = \frac{H_1 + H_2}{FC \times CV} \times 100$$

$$\text{Power Rating } P_{th} = \frac{FC \times CV \times \eta_{th}}{3600 \times 100}$$

2.6 Greenhouse Gas Estimation

Annual CO_2 emission from the fuel saving was estimated with the help of generalized approach as proposed by North Carolina Division of Air Quality[16].

$$\text{Emission}^{CO2} = \text{Fuel} \times CO_2 \text{ emission factor}$$

The CO_2 emission on annual basis is dependent on the type of fuel and its consumption per annum. The CO_2 emission factor for woody fuel was taken as 1.59 kg CO_2/kg.

2.7 Thermal Performance of Single Pot Improved Cookstove

The observation taken during testing of single pot improved cookstove is presented in Table 2.4. Total fuel consumed was 2.2 kg in two hours duration. The stove in working is illustrated in Fig.2.4.

Table 2.4: Test Result of Water Boiling Test of Single Pot Improved Cookstove

S.No.	Particular	Value
1.	Time of starting test (t_1)	11:15 am
2.	Time of completing the test (t_2)	01:15 pm
3.	Initial volume of water (W_1)	5 lit.
4.	Final volume of water (W_f)	2.95 lit
5.	Amount of water evaporated (w_1)	2.05 lit
6.	Initial temperature of water (T_{iw})	18 °C
7.	Final temperature of water (T_{fw})	98 °C
8.	Mass of fuel consumed	2.0 kg

Fig. 2.4: Single Pot Improved Cookstove in Operation

Thermal efficiency

$$\eta_{th} = \frac{5\times4.187\times(98-18)+2260\times2.05}{2\times15491}\times100$$

$= 20.36\ \%$

Power rating

$$P_{th} = \frac{2\times15491\times20.36}{3600\times100}$$

$= 1.75\ kW$

Thermal efficiency of single pot improved cookstove was estimated about 20.36 % and its power rating is about 1.75 kW.

2.8 Thermal Performance of Double Pot Improved Cookstove

The observation taken during testing of double pot improved cookstove is presented in Table 2.5. Total fuel consumed was 2.2 kg in two hours duration. The stove in working is illustrated in Fig.2.5.

Table 2.5: Performance of Double Pot Improved Cookstove

S.No.	Parameters	Value
1.	Time of starting test (t_1)	10:45 am
2.	Time of completing the test (t_2)	12:45 pm
3.	Initial temperature of water (T_1)	18 °C
4.	Final water temperature in first pot (T_2)	98 °C
5.	Final water temperature in second pot (T_3)	98 °C
6.	Initial volume of water in first pot (W_1)	5.0 lit.
7.	Initial volume of water in first pot (W_2)	3.0 lit.
8.	Volume of water evaporation in first pot(w_1)	3.05 lit
9.	Volume of water evaporation in second pot(w_1)	2.60 lit
10.	Total initial volume of water (W_i)	8.0 lit.
11.	Total final volume of water (W_f)	5.65 lit.
12.	Mass of fuel consumed	2.2 kg

Thermal efficiency

$$\eta_{th} = \frac{5\times4.187\times(98-18)+3\times4.187\times(98-18)+2260\times(8-5.65)}{2.2\times15491}\times100$$

$= 23.45\ \%$

Power rating

$$P_{th} = \frac{2.2\times15491\times23.45}{3600\times100}$$

$= 2.21\ kW$

Thermal efficiency of double pot improved cookstove was estimated about 23.45 % and its power rating is about 2.21 kW.

Fig. 2.5: Double Pot Improved Cookstove in Operation

2.9 Exergy Assessment of Improved Cookstove

Ultimate analysis of Desi babul (*Acacia nilotica*) used for thermal testing of improved cookstove is presented in Table 2.6.

Table 2.6: Ultimate Analysis of Biomass (wt %)

Biomass	C	H	O	N
Acacia nilotica	48.82	4.78	0.28	46.12

Higher heating value of unit biomass

$$HHV = [33.5(48.82) + 142.3(4.78) - 15.4\,(0.28) - 14.5\,(46.12)]10^{-2}$$

$$=16.42MJ$$

Lower heating value of unit biomass

$$LHV = HHV\,(1 - 0.056) - 2.440[(0.056 + 9(0.0478)]$$

$$= 14.32MJ$$

Exergy of biomass

$$Ex_{bioamass} = \beta\ LHV$$

Where β is the quality factor and can be calculated as follows:

$$\beta = \frac{1.0412 + 0.2160\left(\frac{4.78}{48.82}\right) - 0.2499\left(\frac{0.28}{48.82}\right)\left[1 + 0.7884\left(\frac{4.78}{48.82}\right)\right] + 0.0450\left(\frac{46.12}{48.82}\right)}{1 - 0.3035\left(\frac{0.28}{48.82}\right)}$$

$\beta = 1.10$

$Ex_{biomass} = 1.10 \times 14.32$

$Ex_{biomass} = 15.77MJ$

Exergy output of single pot improved cookstove

$$\left(Ex_{out}\right)_{Single\ pot} = (5 \times 4.187 \times (371 - 291) + 2260 \times 2.05)\left[1 - \frac{295}{371}\right]$$

(Ex_{out}) single pot = 1.278 MJ

Therefore, exergy efficiency can be written as follows:

$$\psi_{single\ pot} = \frac{1.278}{15.77} \times 100 = 8.103\ \%$$

Exergy output of double pot improved cookstove

$$\left(Ex_{out}\right)_{double\ pot} = 5 \times 4.187 \times (371 - 291) + 3 \times 4.187 \times (371 - 291) + 2260 \times (8 - 5.65)\left[1 - \frac{295}{371}\right]$$

(Ex_{out}) double pot = 1.66 MJ

Therefore, exergy efficiency can be written as follows:

$$\psi_{double\ pot} = \frac{1.66}{15.77} \times 100 = 10.52\ \%$$

The energy and exergy efficiency of both single and double pot improved cookstoves is illustrated in Fig.2.6. It reveals that the exergy efficiency of single pot cookstove is calculated about 8.10 % whereas in double pot cookstove its values estimated about 10.52 %. Thermal efficiency of double pot improved cookstove also found higher than that of single pot cookstove. It also reveals the better fuel saving compared to single pot cookstove.

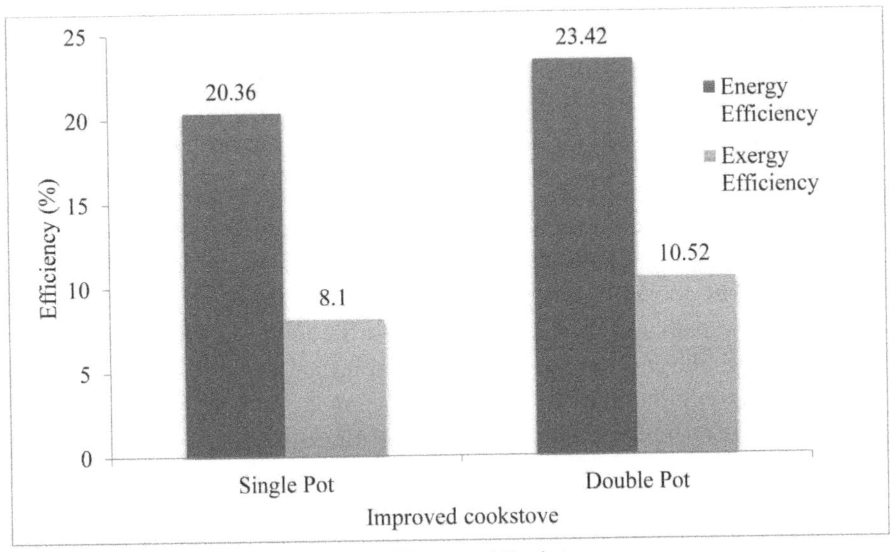

Fig. 2.6: Energy and Exergy Efficiency of Improved Cookstove

2.10 Environmental Benefits of Improved Cookstove

The tradition cookstove has thermal efficiency around 10 percent and consume about 5 kg of fuel per day to prepare meals twice in day. The single pot improved cookstove normally prefer in rural areas where they can save approximately 2.45 kg of wood per day. Hence 928 kg of fuel wood saving per annum can be made possible whereas, double pot cookstove is in position to save about 1050 kg of fuel wood per annum. Similarly, the designed wood gas is capable to save 1135 kg of fuel wood on annual basis. As per as CO_2 reduction is concern single pot, double pot and wood gas stove can save about 1476 kg, 1670 kg and 1805 kg of CO_2 per annum by saving the fuel wood as presented in Table 2.7. Durable cookstove having chimney and it help to expel the smoke produced during combustion and keep the kitchen clean. The designed wood gas stove has very good potential to save fuel wood but operation precaution is needed. It should be operated in open condition or properly ventilated place.

Table 2.7: Fuel Saving and CO_2 Emission Potential

Type of Cookstove	Fuel save per annum (kg)	CO_2 emission (kg per annum)
Single pot	928	1476
Double pot	1050	1670
Wood gas stove	1135	1805

2.11 Design and Development of Wood Gas Stove

In addition to these two improved cookstove, a new stove which work on gasification principle was also developed by MPUAT, Udaipur. It is essentially a inverted down draft type gasifier stove. The system was designed in following manner:

a. **Energy Needed (Q_n)** – This is the amount of heat to be supplied by stove to cook the food. This amount of energy need is depends on the amount of food to be cooked. Mean per capita energy consumption was estimated by Ravindranath and Ramakrishna[17] with different type of fuels and cooking devices. The energy requirement per capita per meal is 3.6 MJ as reported for firewood three stone fire. Hence, the amount of energy needed to cook food in one hour for a family of four members is about 14.4 MJ.

$$Q_n = 14.4\ MJ\ h^{-1}$$

b. **Fuel consumption Rate (FCR)**– The amount of fuel is required to supply sufficient heat energy to cook the food. The stove is working on gasification principle with the assumption that gasification efficiency (η_g) is about 65 %. This can be computed as follows:

$$FCR = \frac{Q_n}{CV \times \eta_g}$$

$$= \frac{14.4}{15.491 \times 0.65}$$

$$= 1.43\text{kJ h}^{-1}$$

c. **Reactor Diameter** – The diameter is a function of specific gasification rate (SGR) and fuel consumption rate (FCR). The values of SGR depends on the gasification mode here natural type of gasification and it values lies between 80-110 kg m^{-2} was taken. Therefore, the diameter of stove can be computed as follows:

$$FCR = SGR \times \frac{\pi}{4} D^2$$

$$D = \left[\frac{1.27\,FCR}{SGR}\right]^{\frac{1}{2}}$$

$$D = \left[\frac{1.27 \times 1.43}{90}\right]^{\frac{1}{2}}$$

$$= 0.1420\text{m}$$

The diameter of gas stove were taken 15 cm

d. **Height of the Reactor** – The stove is batch type and height required to hold the fuel is to be calculated as follows:

$$H = \frac{SGR \times T}{\rho_{fuel}}$$

$$= \frac{90 \times 1}{350}$$

$$\cong 0.26m$$

The height of designed gas stove was found 26 cm. The actual height of the system was kept 38 cm to accumulate the proper mixing of secondary air for proper combustion of the producer gas. The schematic line diagram of designed stove is presented in the Fig.2.7

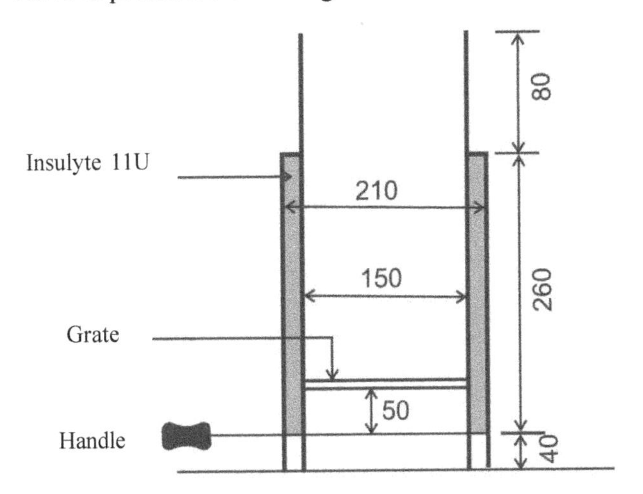

Fig. 2.7: Schematic of Wood Gas Stove

The insulation was made inside the stove to minimize heat loss, MS anchor was welded inside reactor to retain the insulation material. The insulyte 11 U was used for insulation. Proper queering was made before actual use.

2.11.1 Thermal Performance of Wood Gas Stove

The stove is filled with biomassto within 10 cm of the top. It is desirable to have the fire spread rapidly laterally across the surface to provide gas over the whole area, therefore, so an easily combustible material, such as diesel soaked cotton jute, grease etc. is placed on top. Once it ignite from top it gives combustible gas which is called producer gas and blue flame was established in 5 minutes and ran continuously for 50 minutes as illustrated in Fig 2.8. The observation taken during testing is presented in Table 2.8.

Fig.2.8: Wood Gas Stove in Operation

Table 2.8: Performance of Wood Gas Cookstove

S.No.	Particular	Value
1.	Time of starting test (t_1)	11:00 am
2.	Time of completing the test (t_2)	11:55 pm
3.	Initial volume of water (W_1)	5 lit.
4.	Final volume of water (W_f)	2.10 lit
5.	Amount of water evaporated (w_1)	2.90 lit
6.	Initial temperature of water (T_1)	19 °C
7.	Final temperature of water (T_2)	98 °C
8.	Mass of fuel consumed	1.0 kg

Thermal efficiency

$$\eta_{th} = \frac{5\times4.187\times(98-19)+2260\times1.08}{1\times15491}\times100$$

$$= 26.43\%$$

Power rating

$$P_{th} = \frac{1\times15491\times26.50}{3000\times100}$$

$$\cong 1.36 \text{ kW}$$

The developed wood gas stove having 26.43 % thermal efficiency and it is closed to 27% as reported by Bhattacharya[18], and its power rating is calculated about 1.36 kW. During the testing the outer surface temperature was recorded about 105 °C, hence there is still chance to minimize conduction and radiation heat losses. The flame temperature was recorded by K type thermocouple and it was about 736 °C. the total cost of construction of wood gas stove is appear around Rs. 1550 and component wise cost of stove is presented in Table 2.9

Table 2.9: Cost Estimates of Wood Gas Cookstove

S.No.	Material	Cost (Rs.)
1.	GI Sheet	210.00
2.	MS Sheet	420.00
3.	Insulyte	280.00
4.	MS bar for grate	110.00
5.	Miscellaneous	240.00
6.	Labour	290.00
	Total	1550.00

2.12 Eco Friendly Biomass Fired Water Heating System

There are a number of water heating systems have been developed and commercially available for domestic water heating. The commercially available water heating systems are working either on solar energy or electricity. In the morning hours, solar radiations are not available and at the same time electricity may not be available in rural areas to heat water. Electricity based water heating system is a costly affair for rural people, whereas solar water heating system also having high initial cost. With these limitations biomass is good option for water heating in rural areas because it is easily available and having an affordable price. Crop residues like maize cob, cotton stalk, sesamum stalk, etc, are good option to use for generating hot water.

Water heating through the burning of biomass is accomplished by either direct combustion or gasification. Traditional method, i.e. direct combustion for water heating consists of an open chulha or three stone heating systems. It gives incomplete combustion, and poor heat transfer efficiency, use of these fuels has a negative impact on the health of household members, especially women and children, when burned indoors without a proper stove to help control the generation of smoke or a chimney to vent the smoke outside[19]. In traditional chulha, smoke emission and poor heat transfer efficiencies are the main reasons which cause more consumption of fuel.

2.12.1 System Description

The heat transfer from the combustion chamber to the water rises with the increase in the contact area between the water and combustion chamber, as reported by Still[20].

Therefore, the developed biomass-fired water heating system consists of two concentric cylinders; i.e. the inner cylinder and the outer cylinder made of SS 304. Actual combustion takes place in the inner cylinder, which has a diameter of 20 cm, whereas the outer shell has a diameter of 30 cm. Within the experiment, the water was poured between these two cylinders. The outer shell was insulated with glass wool to minimize heat loss. The grate was made of a mild steel round bar. To maintain the proper draft during the combustion of the biomass, a chimney was placed on the top of the combustion chamber, as illustrated in Fig.2.9. One water tap was placed at on the upper side of the water tank to drain hot water, and another tap was placed at the bottom side to drain the water when the system is not in use. The technical specification of a developed water heating system is presented in Table 2.10.

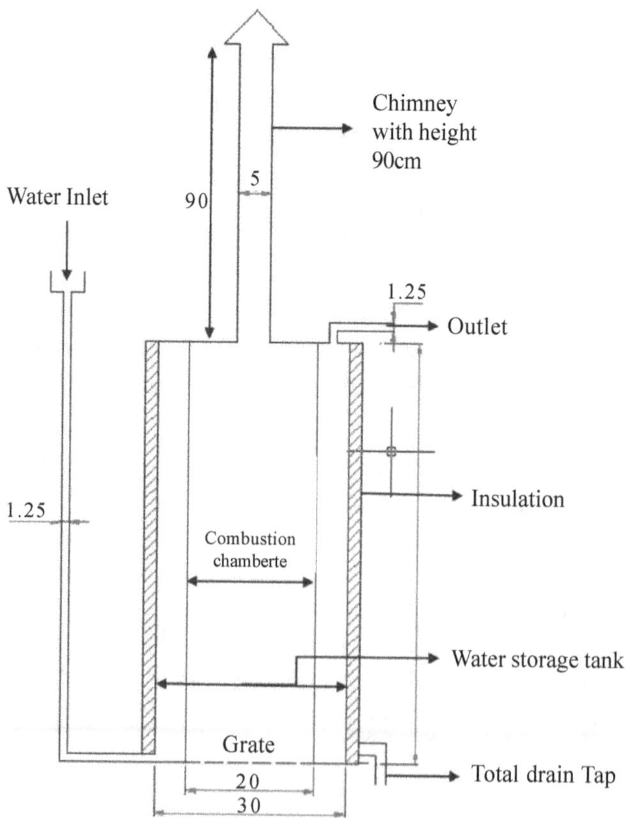

Fig. 2.9: Schematic of Biomass Fired Water Heating System (all dimension in cm).

Table 2.10: Technical Specification of Biomass Fired Water Heating System.

S.No.	Component	Dimension in cm		Material
1.	Inner cylinder	Diameter	= 20	SS 304
		Height	= 60	
2.	Outer cylinder	Diameter	= 30	SS 304
		Height	= 60	
3.	Chimney	Diameter	= 5	GI
		Height	= 90	
4.	Grate	Diameter	= 19	MS round bar
5.	Insulation	Thickness	= 0.4	Glass wool
6.	Insulation cover	Diameter	= 30.8	Aluminium sheet
		Height	= 90	

2.12.2 Thermal Performance of Developed System

Thermal performance of biomass fired water heating system was carried out with two different woody biomass, i.e. Neem (*Azadirachta Indica*) wood and Desi Babul (*Acacia nilotica*) wood. In the initial inception water was filled with rated water holding capacity i.e., 25 litre and combustion chamber was filled with 1.0 kg of fuel wood. Initial water temperature was 30 °C. As fuel wood started burning, water temperature inside the heater was recorded. When temperature gradient was observed (≈ 30 °C), water drain tap was open to collect hot water. Six replications of each of wood test were performed to achieve maximum accuracy.

Thermal performance with Neem (Azadirachta Indica) Wood

The developed system was tested with Neem (*Azadirachta Indica*) wood with 10% moisture level. Temperature profile with different water discharge batches for Neem wood test is shown in Fig. 2.10.

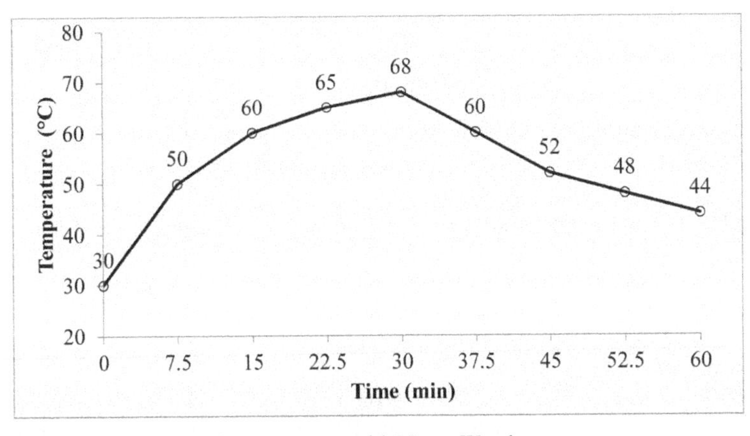

Fig. 2.10: Variation of Temperature with Neem Wood

Fig. 2.10 reveals that the first batch of hot water of 10 litres was discharged at 50°C in 7.5 minutes. Similarly, system delivered 8 batches of 10 litres of hot water, equal to 80 litres per hour, by burning 1 kg of wood. Maximum temperature was recorded at 68°C in this test at discharge of 4th batch. Subsequently, it decreased gradually, and last batch of 10 litres of water was delivered at 44°C. Thermal efficiency of water heater was estimated at about 52.5%.

2.12.6 Thermal Performance with Babool (Acacia nilotica) Wood

The developed system was tested in a similar fashion as with neem wood. The temperature profile with different water discharge batches using babool wood as fuel is presented in Fig. 2.11, which reveals that the first 10 litre batch of hot water was discharged at a temperature of 51°C in 7.5 minutes. Maximum temperature was recorded as approximately 76 °C in the first babool wood test, then at the discharge of the 4th batch it decreases gradually to deliver the last batch of 10 litres of hot water at 50°C. In this test, the maximum temperature attained and the delivery temperature of the last 10 litre batch of hot water both were higher than with the neem wood test. The thermal efficiency of the developed system with babool wood was found to 54.8%.

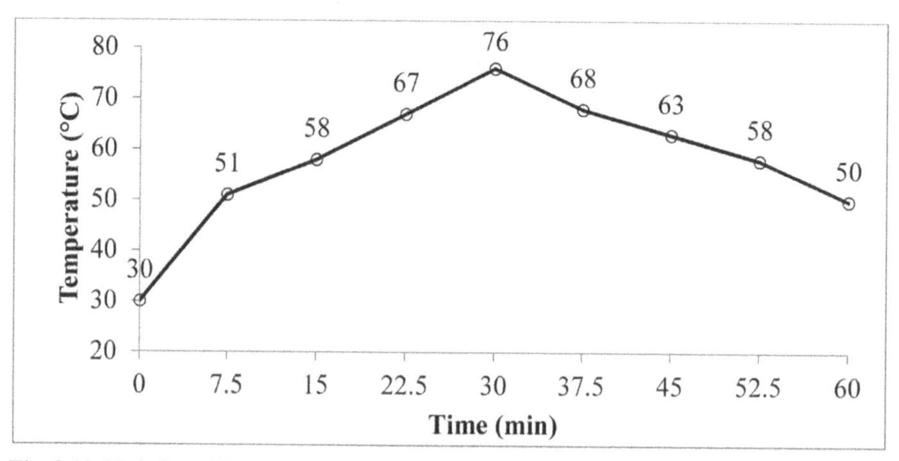

Fig. 2.11: Variation of Temperature with Time when tested with Babool wood

References

[1]. Judy L, Trimble JL, Van Hook RI. Biomass for Energy: the Environmental Issues. Biomass 1984; 6: 3-13.

[2]. Xu R, Ferrante L, Briens C, Berruti F. Flash pyrolysis of grape residues into biofuel in a bubbling fluid bed. J. Anal. Appl. Pyrol. 2009: 86, 58–65.

[3]. Kobayashi N, Fan L-S. Biomass direct chemical looping process: A perspective. Biomass amd Bioenergy, 2011;35:1252-1262.

[4]. Hall DO, Rosillo-Calle F, William RH, Wood J. Biomass energy supply and prospect. In Johanssaon TB. Kelly H. Reddy AKN, Williams RH editors. Renewable Energy Sources for fuel and electricity. Washington DC: Island press 1993: 593-651.

[5]. Bhattacharya SC, Salam Abdul P. Low greenhouse gas biomass option for cooking in the developing countries. Biomass and Bioenergy 2002; 22: 305-317.
[6]. Faaji A. Modern Biomass Conversion Technologies. Mitigation and Adaptation Strategies for Global Change 2006; 11: 343–375.
[7]. Smith KR, Uma R, Kishore VVN, Zhang J, Joshi V, Khalil MAK. Greenhouse implications of household stoves: an analysis for India. Annu. Rev. Energy Environment 2000a; 25:741–63.
[8]. Smith KR, Samet JM, Romieu I, Bruce N. Indoor air pollution in developing countries and acute lower respiratory infec-tions in children. Thorax 2000b;55: 518–32.
[9]. Hossain MMG. Improved cookstove and biogas programmes in Bangladesh. Energy for Sustainable Development 2003;7:97–100.
[10]. Koyunc T , Pinar Y. The emission from a space heating biomass stove. Biomass and Bioenergy 2007; 31: 73-79.
[11]. Karekezi S, Ranja T. Renewable Energy Technologies in Africa. Zed Books, London 1997
[12]. Karekezi S, Teferra M., Mapako M . SPECIAL ISSUE - Africa: Improving modern energy services for the poor. Energy Policy, 2002b; 30: 1015-1028
[13]. Bilgen S, Kaygusuz K, Sari A. Renewable energy for a clean and sustainable future. Energy Sources, Part A: Recovery, Utilization, and Environmental Effects 2004;26(12):1119–29.
[14]. Finet C. Heating value of municipal solid waste', Waste Management and Research 1987; 5: 141–145.
[15]. Szargut J, Styrylska T. Approximate evaluation of the exergy of fuels. Brennstoff Wärme Kraft 1964; 16: 589-596.
[16]. North Carolina Division of Air Quality report on Greenhouse gas emission guidelines: stationary combustion sources. Available on http://www.ncair.org/monitor/eminv/forms/Stationary_Combustion_Sources.pdf (retrieve on June, 2012)
[17]. Ravindranath NH, Ramakrishna J. Energy options for cooking in India. Energy Policy 1997; 25: 63-75.
[18]. Bhattacharya SC, Attalage RA, Augustus Leon M, Amur GQ, Abdul Salam P, Thanawat C. Potential of biomass fuel conservation in selected Asian countries. Energy Conversion & Management 1999;40:1141–62.
[19]. Ciunel K, Radziemska EK. Utilization of rapeseed pellet from fatty acid methyl ester production as an energy source. Environment Technology 2014; 35(2):195-201.
[20]. Still D, Hatfield, M, Scott, P. Capturing heat two, Part 2: Simple water heater. Aprovecho research centre. (http://www.aprovecho.org/lab/rad/rl/stove-design/category/1 assessed on March 15, 2015)

3

Biogas Technology

3.1 Introduction

Nature has a provision for destroying and disposing off wastes and dead plants and animals. This decay or decomposition is carried out by tiny micro-organisms called bacteria. Making of farm-yard manure (FYM) and compost is also through decomposition of organic matter (OM). When a heap of vegetable or animal waste and weeds etc., die or decompose at the bottom of backwater or shallow lagoons, bubbles can be noticed rising to the surface of water. Sometimes these bubbles burn with dancing flame at dusk. This phenomenon has puzzled man for a long time. It was only during the past hundred years that Scientists unlocked this secret as the decomposition process. The gas thus produced was and is still called "Marsh Gas." The technology of harnessing this gas under artificially created conditions is known as Biogas Technology.

3.2 Biogas Technology

Biogas Technology has a very significant role to play in integrated agricultural operations, rural sanitation, large-scale dairy farms & sewage disposal etc. It is estimated that cattle dung, when passed through a Biogas unit, yields 30-40% more net energy and about 35-45% more Nitrogen in manure as compared with that obtained by burning dung cakes and ordinarily prepared compost, respectively. Besides, from a biogas plant both the products are obtained. There are about 250 million bovine (cattle and buffalo) population in India and one biogas unit for small family requires about 3-5 cattle heads, thus about 10 million family size plants fed on cattle and buffalo dung can be installed.

3.2.1 Process Description

Biogas generation is a process widely occurring in nature and can be described as a biological process in which biomass or organic matter, in the absence of Oxygen, is converted into Methane and Carbon dioxide. It is characterized by low nutrient requirement, and high degree of waste stabilization process where biogas is one of the two useful products; the other being enriched organic manure in the form of digested slurry. It is essentially a three stage process involving following reactions:

1) Hydrolysis 2) Acid formation and 3) Methane generation.

For all practical purposes the first two steps are often defined as a single stage, i.e. hydrolysis and acid formation stages are grouped as acid formation stage. Micro-organisms taking part in this phase are termed as acid formers. As a group, these organisms are rapidly growing and are not much dependent upon surroundings. Products of first two stages serve as the raw material for the third stage where organic acids are utilized as carbon source by Methane forming micro-organisms, which are also known as Methanogens. These Methanogens are more susceptible to their surroundings. The tolerated pH range is 6.8 to 7.5 with optimum 7.0. Any departure from this range is inhibitory. Atmospheric Oxygen is extremely toxic for methanogens, as they are strict anaerobes.

3.3 Parameters Affecting Anaerobic Digestion

There are several parameters which affect the anaerobic digestion / gas yields and they can be divided into two parts:

3.3.1 Environmental Factors

There are few environmental factors which limit the reactions if they differ significantly from their optimum levels. Factors of most interest are (a) temperature, (b) pH and (c) nutrient contents of the raw materials,

(a) Temperature

It is a factor which affects most small & medium size biogas installations in developing countries. There are three zones of temperature in which biogas is produced by anaerobic fermentation of organic matter, viz.: 1) Mesophillic, 2) Thermophillic and 3) Psycrophillic zones. The optimum temperature of digester slurry in Mesophillic zone is 35°C, 55°C in Thermophillic zone and 10°C in Psycrophillic zone. In different temperature zones different sets of microbes, (bacteria) especially the methanogens remain active; whereas the other two groups of microbes either remain dormant and thus more or less inactive as far as the anaerobic digestion is concerned or get killed. However, the rate of fermentation is much faster at high temperature. Most rural household biogas plants (digesters) in developing countries operate at ambient temperatures, thus digester slurry temperature is susceptible to seasonal variation but is more dependent on the ground temperature than the atmospheric temperature. As a result, gas output in winter falls by up to 50 %. Below a slurry temperature of

10°C all the reactions cease to take place but revive gradually with the rise in temperature.

(b) pH

The pH range suitable for gas production is rather narrow i.e. 6.6 to 7.5. Below 6.2 it becomes toxic. pH is also controlled by natural buffering effect of NH_4^+ and HCO_3^-ions. pH falls with the production of volatile fatty acids (VFAs) but attains a more or less constant level once the reaction progress.

(c) Nutrient Concentration

Biogas producing raw materials can be divided into two parts i.e. 1) Nitrogen rich and 2) Nitrogen poor. Nitrogen concentration is considered with respect to carbon contents of the raw materials and it is often depicted in terms of C to N ratio. Optimum C/N ratio is in the range of 25 to 30:1. In the case of cattle dung the problem of nutrient concentration does not exist as C/N ratio is usually around 25:1.

3.3.2 Operational Factors

Operational factors contributing to the gas production process are: (a) retention time (RT) - also referred as detention or residence time, (b) slurry concentration and (c) mixing.

(a) Hydraulic Retention Time (HRT)

The number of days the feed material is required to remain in the digester to begin gas production – is the most important factor in determining the volume of the digester which in turn determines the cost of the plant; the larger the retention period, higher the construction cost. In India, the different HRTs are recommended for three different temperature zones as shown in Table 3.1.

Table 3.1: Different HRT Zones

Zone	Average ambient temperature	HRT (days)	Recommended regions
I	>20°C	30	Andhra Pradesh, Goa, Karnataka, Ketala, Maharashtra, Tamil Nadu, Pondicherry and Andaman & Nicobar Islands
II	15-20°C	40	Bihar, Gujrat,Haryana, Jammu region of J&K, Madhya Pradesh, Orrisa, Punjab, Rajasthan, Uttar Pradesh and West Bengal
III	<15°C	55	Himanchal Pradesh, North-eastern states, Sikkim, Kashmir region of J&K, and hill districts of UP

(b) Slurry Concentration

This is denoted by dry matter concentration of the inputs. The optimum level for cattle dung slurry in the range of 8 to 10% and any variations from this result in lower gas output. Mixing four parts of dung with five parts of water forms a slurry with dry matter concentration of about 9%, whereas 1 part of dung to 1 part to water would give a slurry concentration of 10%. This also affects the loading rate etc.

(c) Mixing & Stirring

Proper mixing of manure to form an homogenous slurry before it is fed in the digester, is an essential operation for better efficiency of biogas systems; whereas proper stirring of digester slurry ensures repeated contact of microbes with substrate and results in the utilization of total contents of the digesters. An extremely important function of stirring is the prevention of formation of scum layer on the upper surface of the digester slurry which, if formed, reduces the effective digester volume and restricts the upward flow of gas to the gas storage chamber. Mixing results in premature discharge of some of the input and a perfectly unmixed system is likely to result in better reaction rate but for the problem of scum formation.

3.4 Constituents

The gas thus produced by the above process in a bio-gas plant does not contain pure methane and has several impurities. A typical composition of such gas obtained from the process is as given in Table: 3.2

Table 3.2: Constitutes of Biogas

S.No	Gases	% (Percentage)
1.	Methane	60.0%
2.	Carbon-dioxide	38.0%
3.	Nitrogen	0.8%
4.	Hydrogen	0.7%
5.	Carbon-monoxide	0.2%
6.	Oxygen	0.1%
7.	Hydrogen Sulphide	0.2%

The calorific value of methane is 35.16 MJ / m^3 and that of the above mixture is about 19.73 MJ/ m^3. However, the bio-gas gives a useful heat of 12.56 MJ/ m^3.

3.5 Calorific value

The calorific values or heat values indicate (Table3.3) that bio-gas can perform works similar to fossil oil in domestic cooking, lighting etc., with better efficiency depending upon the methane content in it. The bio-gas has also the potential for use in internal combustion engines used for pumping water etc. for which research and development works are in progress. Biogas, therefore, has a bright future as an alternate renewable source of energy for domestic and farm use.

Table.3.3: Calorific Value of Different Fuel

Commonly used fuels	Calorific values in MJ	Thermal efficiency
Bio-gas	19.73/m^3	60%
Dung cake	8.76/kg	11%
Firewood	20.84/kg	17.3%
Diesel (HSD)	44.17/kg	66%
Kerosene	45.42/kg	50%
Petrol	46.05/kg	—

3.6. Pre-requisites of Bio-Gas System

(i) **Land and Site:** While selecting a site for a bio-gas plant, following aspects should be considered:

a) The land should be leveled and at a higher elevation than the surroundings to avoid runoff water.

b) Soil should not be too loose and should have a bearing strength of 2 kg/cm^2

c) It should be nearer to the intended place of gas use.

d) It should also be nearer to the cattle shed/ stable for easy handling of raw materials.

e) The water table should not be very high.

f) Adequate supply of water should be there at the plant site.

g) The plant should get clear sunshine during most part of the day.

h) The plant site should be well ventilated as methane mixed with oxygen is very explosive.

i) A minimum distance of 1.5m should be kept between the plant and any wall or foundation.

j) It should be away from any tree to make it free from failure due to root interference.

k) It should be at least 15m away from any well used for drinking water purpose.

l) There should be adequate space for construction of slurry pits.

(ii) **Feed for Gas Plants:** The feed for gas plants in India mainly comprises of dung from cattle. Although, quantity of dung per cattle depends upon health, age, type and many other factors, it is generally believed that, average cattle yield is about 10 kg dung per day. On this presumption, the number of cattle required for various sizes of gas plants as has also been recommended by KVIC.

(iii) **Temperature:** Temperature plays the most important role in the bio-gas production. The total amount of gas production from a fixed weight of organic waste is best when the temperature is within the messophillic range 25°C-37°C and thermophillic range between 45°C-55°C. The gas yield is maximum in the thermophillic region and the period of digestion is also reduced. It takes about 55 days in messophillic range for digestion where as it takes about 7 days in thermophillic region.

(iv) **Hydrogen ion Concentration:** pH of slurry in the digester should be maintained between 6.8 and 7.2 for optimum gas production and this can be accomplished by maintaining proper feeding rate. pH indicates the acidity and alkalinity of the feed mixture. Any excessiveness of acidity or alkalinity would affect gas production. There are a number of ways to correct the pH if the slurry becomes acidic or alkaline.

(v) **Agitation:** Mechanical agitation of the scum layer and slight stirring of slurry improves gas production but violent stirring retards it.

(vi) **Solid Content:** The solid content in the slurry should be maintained between 7.5 to 10 per cent for optimum gas production.

(vii) **Carbon to Nitrogen Ratio:** A carbon to nitrogen ratio of 20: 1 to 30: 1 is found to be optimum for bio-gas production. Carbon to nitrogen ratio of various materials is given in Table -3.4 as a guide so that the C: N. ratio of biogas feed mixture is kept at desired level.

Table 3.4: Carbon to Nitrogen Ratio of various materials

Sr. No.	Material	Nitrogen Content (%)	Ratio of Carbon to Nitrogen
1.	Urine	15.18	8:1
2.	Cow dung	1.7	25:1
3.	Poultry manure	6.3	N.A.*
4.	Night soil	5.5-6.5	8:1
5.	Grass	4.0	12:1
6.	Sheep waste	3.75	N.A. *
7.	Mustard straw	1.5	20:1
8.	Potato tops	1.5	25:1
9.	Wheat straw	0.3	128:1

3.7. Main Features of the Biogas Plant

On the basis of the gas holder the present biogas plants are classified mainly into two groups -fixed dome type or floating drum type. Both the type of plants have the following functional components:

(i) **Digester :** This is the fermentation tank and is built partially or fully underground. It is generally cylindrical in shape and made up of bricks and cement mortars. It holds the slurry within it for the period of digestion for which it is designed.

(ii) **Gas holder:** This component is meant for holding the gas after it leaves the digester. It may be a floating drum or a fixed dome on the basis of which the plants are broadly classified. The gas connection is taken from the top of this holder to the gas burners or for any other purposes by suitable pipelines. The floating gas holder is made up of mild steel sheets and angle iron and is required to exert pressure of 10 cms of water in the gas dome masonry and exert a pressure upto 1m of water column on the gas.

(iii) **Slurry mixing tank:** This is a tank in which the dung is mixed with water and fed to the digester through an inlet pipe.

(iv) **Outlet tank and slurry pit:** An outlet tank is usually provided in a fixed dome type of plant from where slurry in directly taken to the field or to a slurry pit. In case of a floating drum plant, the slurry is taken to a pit where it can be dried or taken to the field for direct applications.

3.8 Classification of Rural Household Digester

There are three basic methods by which rural household biogas digesters in developing countries are operated in practice, namely: (i) batch, (ii) semi-continuous and (iii) semi-batch digesters.

(a) Batch Digester

In this process the whole digester is filled with raw materials for gas production along with some starting (seed) material. This is allowed to ferment and produce gas over a certain length of time and when gas yields become very low the digester is emptied of all the sludge which can be supplied as manure. In this system gas production begins at a low level and goes on increasing only to drop down again after reaching the peak. Because of variable gas production level, high cost and periodic emptying and filling of digesters, this process has not become popular. Examples of these digesters are small size garbage plant and crop-residues plant.

(b) Semi-Continuous Digester

The rural household digesters are fed once a day and the fresh input displaces the same volume of spent materials from the digester. Every day a certain quantity of fresh inputs is fed into the digesters which is expected to remain in the digester for a prescribed retention time and produces gas over this length of time before being discharged.

(c) Semi-Batch Digester

A combination of batch-fed and semi-continuous fed digestion is known SHF digestion. Such a digestion system is used where the waste like garbage etc., which are available on daily or weekly basis but cannot be reduced to make slurry. In the semi-batch system, the animal manure can be added on a daily basis after the initial loading is done with garbage, agricultural waste, leaves, crop residence and water hyacinth etc.

3.9 Size Selection of Rural Household Biogas Plants

Size of the rural household biogas plant to be installed should be selected on the basis of gas requirement and the livestock manure availability with the beneficiaries. Since cattle dung is the main substance for the biogas plant in rural India, Table 3.5 given below shows the relationship between plant capacity, daily cattle dung requirement and gas use.

Table 3.5: Plant Capacity wise Dung Requirement

Size No	Plant capacity m^3	Daily dung Required (kg)	Approximate No. of cattle	No. of family members
1	1	25	2 – 3	3 - 4
2	2	50	4 – 6	5 - 8
3	3	75	7 – 9	9 - 12
4	4	100	10 – 12	13 - 17
5	6	150	12 – 20	18 - 25

3.10 Popular Designs of Biogas Plant Models

There are three popular Indian designs of biogas plants namely: KVIC, Janata and Deenbandhu Biogas plants. For construction of KVIC & Janata model plants - Indian Standard IS:9478-1986 released by Bureau of Indian Standards should be followed. Brief description of the three models is given below.

(a) KVIC Plant

It was in or around the year 1945 that Scientists at Indian Agricultural Research Institute (IARI), New Delhi worked on semi-continuous flow digesters and in the year 1961 Khadi and Village Industries Commission (KVIC) patented a design which is being popularized by various agencies in many countries. This design consists of a deep well shaped underground digester connected with inlet and outlet pipes at its bottom, which are separated by a partition wall dividing the ¾th of the total height into two parts. A mild steel gas storage drum is inverted over the slurry which goes up and down around a guide pipe with the accumulation and withdrawal of gas. Now FRP and ferro-cement gas holders are also being used in the KVIC plant (Fig. 3.1).

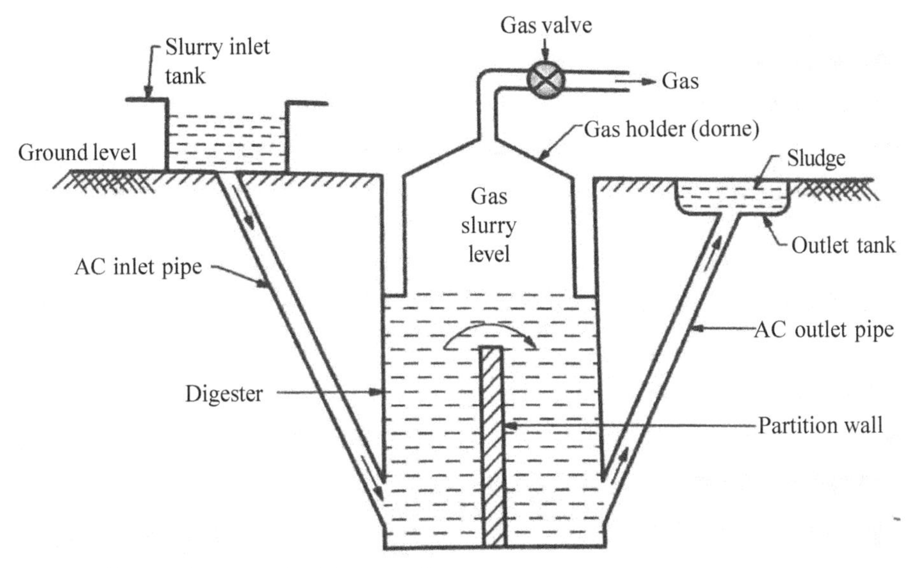

Fig. 3.1: Schematic View of KVIC Biogas Plant

Advantages of Moveable Drum

(i) Constant gas Pressure.

(ii) Minimum gas leakage problem.

(iii) Higher gas production

(iv) Scum problem is minimum.

(v) Pressure is naturally equalizes.

(vi) No danger of mixing between biogas and external air. Hence no danger of explosion.

Disadvantages of Movable Drum Type

(i) Higher cost

(ii) Higher maintenance cost.

(iii) The outlet pipe should be flexibale. It requires regular attention.

(iv) Heat is lost through gas holder.

(b) Janata Plant

The Janata model is a fixed roof biogas plant which was developed by PRAD in 1978. This is also a semi-continuous flow plant. The main feature of the Janata design is that the digester and gas holder are part of a composite unit made of bricks and cement masonry. It has a cylindrical digester with dome shaped roof and large inlet and outlet tanks on two sides. It requires shuttering for making the dome shaped roof and a skilled & trained master mason is a must for the construction of a Janata Biogas plant. This plant costs about 20-30% less than the KVIC floating drum type plant (Fig. 3.2).

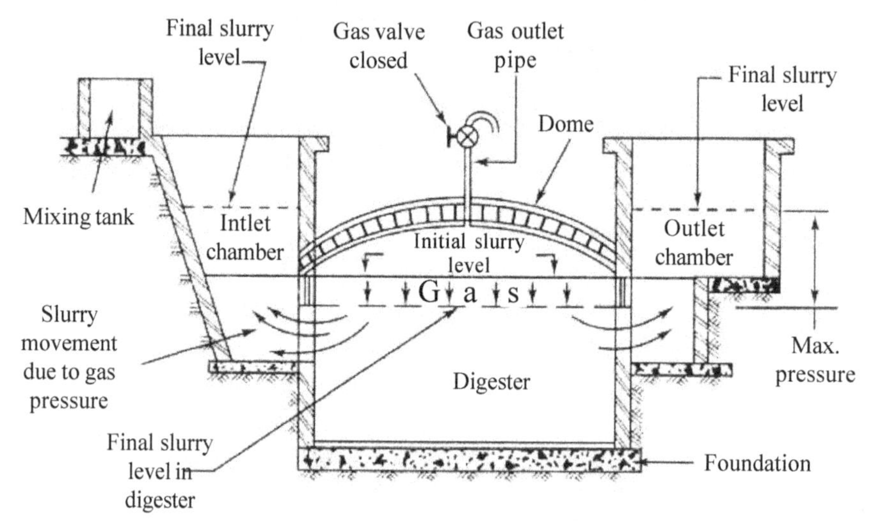

Fig. 3.2: Schematic View of Janata Fixed Dome type Biogas Plant

(c) Deenbandhu Plant

Unlike Janata biogas plants, for constructing plants of this design no shuttering is required for making the dome shaped roof. This also results in less labour and time required for completing the construction.

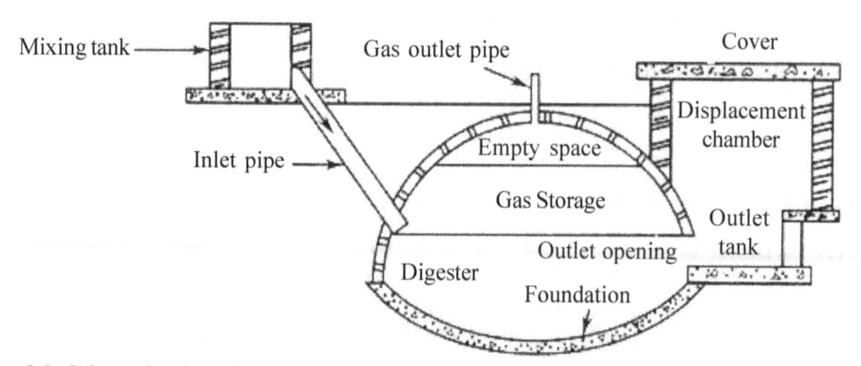

Fig. 3.3: Schematic View of Deenbandhu Type Biogas Plant

3.11 Comparisons among Family Size Biogas Plants

#	KVIC	Janata	Deenbandhu
1.	The digester of this plant is a deep well shaped masonry structure. In plants above 3m^3 there is a partition in the middle of the digester.	Digester of this plant is a shallow well shaped masonry structure. No partition wall is provided.	Digester is made of segments of two spheres one each for the top and bottom.
2.	Gas holder is generally made of mild steel. It is inverted into the digester and goes up and down with formation and: utilization of gas,	Gas holder is integral part of the masonry structure of the plant, Slurry from the gas storage portion is displaced out of the digester with the formation of gas and comes back when it is used.	The structure described above includes digester and the gas storage chamber, Gas is stored in the same way as in the case of Janata plants.
3.	The gas is available at a constant pressure of about 10cm of water.	Gas pressure varies from 0 to 90 cm of water.	Gas pressure varies from 0 to 75 cm of water.
4.	Inlet and outlet connections are provided through A,C pipes	Inlet and outlet tanks are. large masonry structures designed to store the slurry displaced out of the digester.	Inlet connection is through AC pipe. 0utlet tank is large masonry designed to store displaced slurry.
5.	The volume of the gas holder governs gas storage capacity of the plant.	.it is the combined volumes of inlet and outlet chambers, (portions of inlet and outlet tanks above the second step.)	It is the volume of outlet displacement chamber
6.	The floating mild steel gas holder need regular maintenance to prevent corrosion. It has short life.	There is no moving pert and hence no recurring expenditure. It also has e long working life.	There is no moving part and hence no recurring expenditure, It also has a long working life.
7.	Installation cost is very high, A 2 m^3 plant costs over Rs. 36,000.00.	It is cheaper than the KVIC type plants, A 2 m^3 plant costs about Rs.30,000.00.	It is much cheaper then KVIC and Janata type plants. A 2 m^3 ‘plant of this design costs Rs 24,000.00.
8.	Digester can be constructed locally the gas holder needs sophisticated workshop facilities.	A trained mason using locally available materials can build entire plant.	Entire plant can be built by a trained. mason using locally available materials.

3.12 Major Benefits of Installing Biogas Plants

It is estimated by the Ministry of Energy, Govt. of India, that alternative sources of energy like biogas plants, wind mills etc. may reduce the dependence on conventional sources of energy by about 20% by the turn of the century, provided promotional efforts are continued. Presently, the cooking media in rural areas consist of burning dung cake, firewood and to some extent kerosene where it is available easily. The installation of bio-gas plants would directly replace the use of above three and in saving them, following gains would be made:

(i) Nearly 30% of available dung which is burnt and wasted would be recovered as bio-gas plants conserve the dung while producing biogas. Again, the dung after digestion in gas plant preserves more of NPK in the dung solids and cellulose which otherwise gets lost if heaped in the open.

The average NPK content of Farm Yard Manure (FYM) is about 0.5, 0.2 and 0.5 percent respectively and it may be observed that biogas slurry is rich in NPK by more than four times than ordinary dung when converted into FYM. When the country is faced with shortage of fertilizers and has to spend enormous amounts for its import, the application of biogas slurry can replace the chemical fertilizers to a large extent. Bio-gas slurry or FYM not only adds NPK but it proves the soil porosity and texture. These are established benefits.

(ii) Second major benefit is that rural people would gradually stop felling trees. Tree felling bas been identified as one of the major causes of soil erosion and worsening flood situation. Government has started massive afforestation programme to tackle the erosion and flood situation. Continued deforestation has been causing ecological imbalances in the environment in which we live. Biogas plants would be helpful in correcting this situation.

(iii) In rural areas kerosene is used for lighting lantern and cooking in a limited way wherever kerosene supply has been made possible. Whatever quantity is used can be replaced by biogas as it can be used for lighting and cooking. This would reduce the dependence on fossil oil directly and in saving foreign exchange.

(iv) Lastly, the most important social benefit would be that the dung being digested in the digester, there would be no open heap of dung to attract flies, insects and infections. The slurry from digesters can be transported to the farm for application in the soil, thus keeping the environment clean for inhabitation. Also, gas cooking would remove all the health hazards of dung cake or fire wood cooking and would keep the woman folk free from respiratory and eye diseases which are prevalent in the villages. Different fuel equivalent to 1 m^3 biogas is presented in Table 3.12

Table 3.12: Equivalent Quantity of Fuel for 1 m^3 of Biogas

Name of the fuel	Equivalent quantities to 1 m^3 of Biogas
Kerosene	0.620 kg
Firewood	3.474 kg
Cowdung cakes	12.296 kg
Charcoal	1.458 kg
Soft coke	1.605 kg
Butane	0.433 kg
Furnance Oil	0.4171
Coal gas	1.177 m^3
Electricity	4.698 kWh

3.13 Bio-gas Application and Appliances

Cooking and Lighting: The main use of tbe bio-gas in rural areas is cooking. Firewood, crop residues or dung cakes are not available on regular basis and so are other conventional sources of energy like kerosene, electricity or coal etc. The other 'use is gas lamp which glows like any bright lamp. Generally, 15 running meter length (RML) of pipes/tubing are allowed in bio-gas plant schemes along with burners and lamps for schemes financed by banks.

(i) **Dual Fuel Engines:** This is a recent appliance where certain modification of air intake system helps carburation of biogas to run the diesel engines. It is well known that diesel engine has wide application in rural areas from irrigation to any stationary operations and these engines can be converted to dual fuel ones. This dual fuel engine is in a position to make use of about 70% biogas and 30% diesel. The economy of running dual fuel engine by biogas is undisputable but it has certain operational difficulties. The main problem is that while biogas plants are located near the house, the running of the diesel pumpsets are required in the fields. As such it is impractical and technically unsound to provide long gas pipes to connect the gas plant with the diesel engines in the field. Another aspect that prohibits the use of bio-gas in diesel engine is that it requires larger gas plants at least 8-10 cum, so as to enable a low 3 HP diesel engine to run for 4-5 hours a day. General the users prefer the sizes of 2-4 cum gas plants as number of cattles are not many. However, considering large biogas plants hold the key to economical operation of dual fuel engines, in future one may see its wide spread application when many such gas plants come into existence.

(ii) **Refrigerations, Incubators and Water Boilers :** There are other applications of bio-gas such as refrigerators, incubators and water boiler, Experimentation in this regard has been going on in some R & D centres whose results are awaited before they become common application items on commercial level.

(iii) **Research and Development in Application of Biogas:** Unless ways and means are found for diverse application of biogas, it may not receive the appreciation in deserves. Community bio gas plants have been mooted to provide cooking gas, street light and drinking water through pumping. Government of India had proposed during 1983-84 to install 500 community gas plants on pilot basis and study their feasibility for running on cooperative basis. The results may be helpful in spreading similar plants to other areas. Besides, many institutions/individuals in the country have attempted to provide alternative gas plant designs and some 30 designs have been tried so far. Yet the KVIC design and Janata Model are the two which are not only but are recognized by the financing institutions as standard designs and this status would be maintained unless better designs come as replacements. In certain parts of the country use of night soil and connecting latrine to biogas plants have already found its social acceptability. However, in most part of the country it is still not socially accepted and when it finds its social acceptability, it would ensure a regular supply of feed besides improving sanitation in rural areas.

Example 3.1: Design a Biogas plant to supply 100 kWh of electricity per day. The engine and generator efficiency are 25% and 80% respectively. The methane content in biogas is 60 %. Determine the number of cattle, biogas required and greenhouse gas mitigation potential.

Solution

Overall efficiency $= 0.25 \times 0.80 = 0.20$

$$\text{Total Energy Required} = \frac{100}{0.2} = 500 \text{ kWh}$$

$$\text{The Calorific Value of methane is} 35\frac{\text{MJ}}{\text{m}^3}; {}^{35}\!/_{3.6}\text{kWh} = 9.97\text{kWh}$$

But, Biogas contains 60% methane,

Therefore, The calorific value of 1 m^3 biogas $= 9.97 \times 0.6 = 5.98\text{kW/m}^3$

$$\text{The total volume of biogas require to produce } 500\frac{\text{kW}}{\text{day}} = \frac{500}{5.98} = 83.61\, m^3$$

But, biogas production from 1kg of cattle dung $= 0.04\text{m}^3$

$$\text{Therefore, The Cattle dung Required} = \frac{84\ m^3/_{day}}{0.04\ m^3/_{kg}} = 2100\ kg\ /\ day$$

Assume 1 cattle will produce 10 kg of dung/day

$$\text{Therefore, total number of cattle} = \frac{2100}{10} = 210\ \text{cattles}$$

Daily charge of slurry is about 2100 kg per day

for the biogas production we require to add equal amount of the water to maintain 8-10% total solid content

Total daily charge slurry = 2100 dung + 2100 water = 4200 kg slurry

Now, the density of slurry = 1090kg/m^3

Therefore, Daily chargre of slurry = 2100kg dung + 2100 kg water = 4200kg slurry

$$\text{Daily volumeric charge of slurry into the digester (v)} = \frac{4200}{1090} = 3.85\ m^3/_{day}$$

Let us, now consider Hydraulic Retention Time (HRT) = 45 Days

Therefore, volume of the digester = v × HRT = 3.85 × 45 = 173.25m3 ≈ 174m^3

As the 10% extra space for the gas collection

Actual Volume of digester = 174 + (0.1 × 174) = 191.4m^3

Volume of digester = C/A × Height

Considering the Diameter to Height ratio is equal of the digester,

$$V = A \times H$$

$$V = \frac{\pi}{4}\ d^2\ h$$

$$V = \frac{\pi}{4}\ d^3$$

$$d = \sqrt[3]{\frac{191.4 \times 4}{3.14}} = 6.24\ m$$

Dimeter = 6.24 m

Height = 6.24 m

Now, Design the Volume of gas holder = 40% of plant capacity

$=191.4 \times 0.4 = 76.56\ m^3$

In addition biogas plant will reduce CO_2 emission. It is observed that specific Carbon Dioxide (CO_2) emmision of various fuels is quite high as given in Table 3.13.

Table 3.13: Specific carbon dioxide Emmission of values fuels

S.No.	Fuel	Emission in kg CO_2 / kWh
1.	Natural gas	0.2
2.	LPG	0.23
3.	Petrol	0.25
4.	Kerosene	0.26
5.	Diseal	0.27
6.	Coal	0.34

Thus, this 100kWh/day biogas plant will reduce the CO_2 emission
= 20 kg CO_2 / kWh of Natural Gas
= 23 kg CO_2 / kWh of LPG
= 25 kg CO_2 / kWh of Petrol
= 26 kg CO_2 / kWh of Kerosene
= 27 kg CO_2 / kWh of Diesel
= 34 kg CO_2 / kWh of Coal

4

Carbonisation of Biomass

4.1 Introduction

Biomass is referred to as an indirect source of solar energy and considered a source of stored chemical energy. Biomass is renewable organic material derived from plants and animals serving as sources of energy. Biomass produced from agricultural sector is big challenges for all over the world due to its improver disposal. Mitigating greenhouse gas emissions and ensuring adequate global food supplies represent two of the last decade's most difficult challenges. Although global food production has benefitted from chemical fertilizers, environmental problems have emerged as a result of their use. Additionally, overuse of fertilizers can result in hardened soil, decreased soil fertility, polluted air and water, and the release of greenhouse gases. There is an urgent need to find an alternative to chemical fertilizers that, ideally, can be sourced in abundant amounts, promotes global food production, enhances CO_2 capture , and does not affect soil health or damage the environment[1]. To sustain agricultural productivity, it is crucial to maintain adequate levels of organic matter in the soil to preserve its physical, chemical, and biological integrity. Biochar, a pyrogenic black carbon, may play an important role in improving soil health, resulting in higher crop yield and absorbing atmospheric carbon dioxide. Biochar is the most auspicious straw management measures and have highest carbon abatement and economic profit[2].

Biochar is produced by heating biomass at high temperatures (300 – 600 °C) in a closed reactor containing no to partial levels of air. Under these conditions, biomass undergoes thermo-chemical conversion into biochar. Because of its numerous potentials uses in agriculture, energy, and the environment, much attention has been given to biochar in both political and academic areas. Biochar can be used in a variety of applications such as energy production, agriculture, carbon sequestration, waste water treatment, and bio-refinery; additionally, biochar provides an alternative strategy for managing organic waste. These advantages have renewed the interest of agricultural researchers in producing biochar from bio-residues and using the product as a soil amendment.

In India, about 686 million tons gross residues are available on annual basis from agricultural crops, and about 234.5 million tons represent the surplus potential[3]. This shows the availability of enough raw materials for an efficient and ecofriendly biochar production unit. The biomass used for biochar production can be classified as illustrated in Fig. 4.1. Fig. 4.2 summarizes the thermo-chemical conversion routes of biomass, including direct combustion to provide heat, liquid fuel and other elements for thermal and electrical generation.

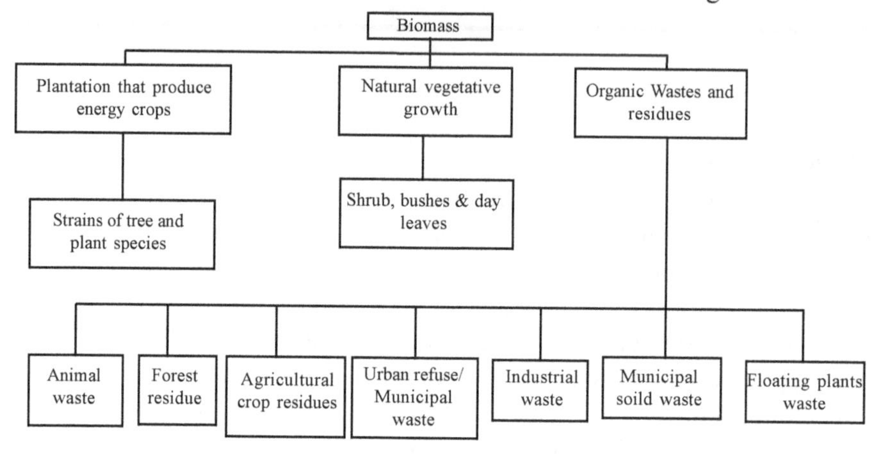

Fig. 4.1: Classification of Biomass

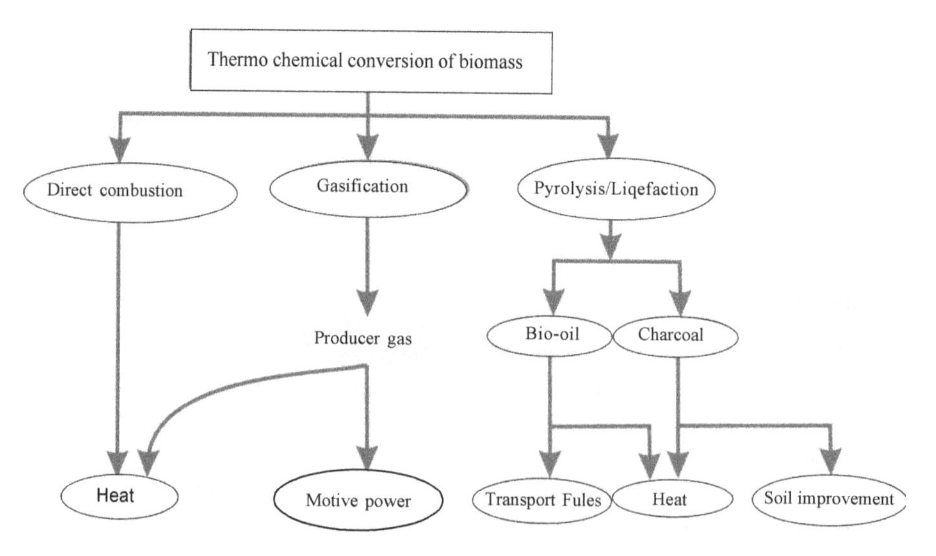

Fig. 4.2: Biomass Conversion Routes

Biochar has acquired new dimensions in the current organic farming era, but its origins are associated with soils of the Amazon region usually referred as "terra-preta" soils, which have been found up to two meters in depth. It is a highly fertile dark-coloured soil that has supported the agricultural needs of the Amazonians for centuries[4]. The concept of Terra-Preta Sanitation (TPS), which has been extensively adopted in Amazon civilization. TSP is a close loop process and very beneficial for a sustainable life style, integrating soil fertility, food security, waste management, and renewable energy. The Terra-Preta Sanitation process is comprised of a with diversion of urine through a charcoal mixture and is based on lactic-acid-fermentation with subsequent vermicomposting. It was found that lacto-fermentation is a biological anaerobic process where no gas and no odor are produced[5].

The presence of terra-preta reveals that humans were deliberately responsible for its creation. Carbonization of biomass for producing biochar has been recorded as long as human evolution has existed[6,7]. Biochar used for soil improvement is produced through a slow pyrolysis process because of its higher biochar yield compared to other production processes[8]. Basically, under a slow pyrolysis process, biomass is heated within the range of 300-600 °C for a longer period[9].

4.2. Biochar and Sustainability

Biochar plays a major role in mitigating climate change, promoting environmental sustainability and increasing agricultural productivity, facilitating soil carbon storage, and improving soil fertility to increase plant and overall crop yield. Lehmann and Joseph[10] have presented four motivational objectives of biochar application, i.e., soil improvement, waste management, climate change mitigation, and energy. Either individually or in combination, these objectives can have either social or financial benefits or both. Biochar always draws attention as a potential input for agriculture, as it can improve soil fertility, aid sustainable production and reduce contamination of streams and groundwater.

Woolf[11] introduced a sustainable biochar concept and reported that the atmospheric CO_2 is utilized by green plants during photosynthesis. Pyrolyzation of the biomass results in bio-oil and biochar. Further, reduction in annual net emission of CO_2, CH_4, and nitrous oxide by 1.8 Pg. Biochar amendment to soil can prevent greenhouse gas emissions from the soil. The biochar can increase the water and nutrient holding capacities of soil, which typically then result in increased plant growth.

4.3. Biochar Production Technologies

Biochar is derived from a wide variety of biomasses including crop residues that have been thermally degraded under different operating conditions. It exhibits a correspondingly immense range of composition. Longer the residence period (up to 4 hours) with moderate temperature (up to 500 °C) biochar yield varied from 15 %– 35 % while bio-oil yield varied between 30% - 50%. On other hands, with lesser residence time (up to 2 second) higher bio-oil (50% - 70%) yield.

Thermochemical processes like pyrolysis and carbonization convert the biomass into bio fuels and other bio energy products. In the pyrolysis process, thermochemical conversion of biomass is carried out in absence of air and at a temperature above 400°C to form a solid product known as biochar. The biochar mainly consists of carbon (C), hydrogen (H), oxygen (O), nitrogen (N), sulfur (S) and ash. Generally, there are three modes of pyrolysis: slow, intermediate and fast. A higher biochar yield was found with a slow pyrolysis process as compared to others[11]. Steiner[12] produced biochar from rice husk using a top-lit updraft gasifier and found that such technology is relatively simple for farmers to produce biochar in the field, with an efficiency of 15–33%. Biochar produced from on-farm available crop residues is sufficient to amend 6.3 – 11.8 % of the production area annually[13].

Carbonization is a slow pyrolysis process that has been in use for thousands of years, and the its main goal is the production of biochar. In slow pyrolysis, a biomass is heated slowly in the absence of air to a relatively low temperature (≈ 400°C) over an extended period of time[14]. Energy can drive the process in the following different ways: (i) directly as a heat of reaction (ii) directly by flue gases from the combustion of feedstock (iii) through indirect heating of the reactor wall using a hot gas; or (iv) through indirect heating of the reactor wall using sand or other non-gas materials. The biochar production process can be classified as illustrated in Fig. 4.3.

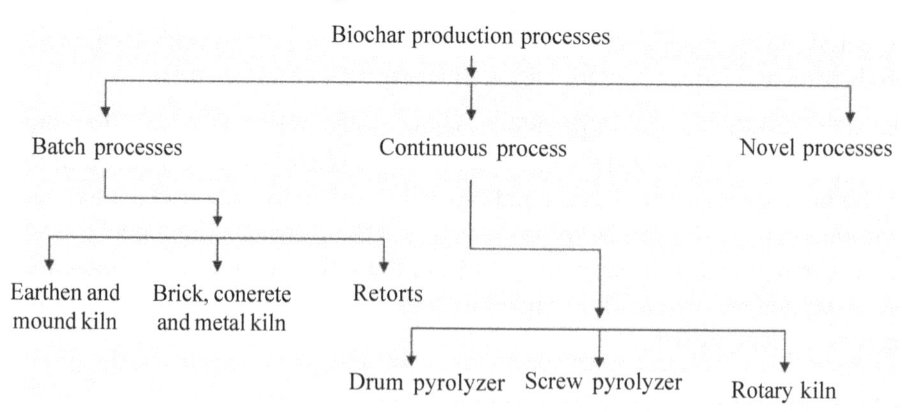

Fig. 4.3: Classification of Biochar Production Processes

4.3.1 Batch Processes

The batch process is an ancient practice and is still practiced in rural areas for biochar production. Though the charcoal yield in such a process varies over the low range of 12.5 to 30 %, it is still preferred in the countryside because of its low operational and construction cost. Batch process for biochar production includes,

(a) Earthen and mound kiln,

(b) Brick, concrete, metal kiln

(c) Retorts

4.3.1.1 Earthen and Mound Kiln

Duku[15] conducted an experiment on the production of biochar by using an earthen mound kiln in Ghana. During the study, they used wood as a feedstock and found that the ground acts as an insulating material that resisted the entry of oxygen during the carbonization process. Masek[16] performed an experiment on biochar production using an earthen mound kilns and found a yield of more than 10%. Bailis[17] used wood as a feedstock for charcoal production and found that moisture content affects the yield of charcoal in traditional processes and found yield of charcoal ranges from 10% to 30 % when using wood as a feedstock. Lohri[18] estimated an average emission component in the case of an earthen mound kilns. His analysis showed that the emission of carbon dioxide (CO_2) into the atmosphere as around 334 ppm. Significant emissions of the products of incomplete combustion adversely affect human health and the environment. FAO[19] found that a higher level of efficiency and product quality could be obtained at a maximum carbonizing temperature of approximately 500° C.

4.3.1.2 Brick, Concrete and Metal Kiln

Kristofferson[20] constructed a rectangular kiln using either masonry blocks or poured concrete combined with steel reinforcements. They, reported charcoal production cycles in a cold climate to be 25-30 days long and in a warm climate to be 33 days with maximum charcoal yields of 25-33%. Deal[21] conducted an experiment to produce biochar using a metallic kiln. During the experiments, five different feed stocks were used: eucalyptus, maize, rice husks, coffee husks, and groundnut shells. The maximum temperatures reached inside the kiln were recorded as being between 400 °C and 600 °C at the top of the kiln and between 600 °C and 800 °C at the bottom. Further, biochar yields were 140-290 g kg^{-1} of the initial biomass weight for eucalyptus, 240-250 g kg^{-1} for maize cobs, 450-490 g kg^{-1} for rice husks, 360-430 g kg^{-1} for coffee husks, and 290-320 g kg^{-1} for groundnut shells.

Pennise[22] tested a Brazilian round brick kiln with a capacity of 20,000 kg of woody biomass and noted a that charcoal yield of approximately 68.9 % with a carbon content of 85.7% and a calorific value of approximately 29.20 kJ/g.

Mwampamba[23] used wood as a feedstock for charcoal production, using a brick and metal kiln and found that the efficiency of production fell between 25 and 35 %. Kammen[24] compared the efficiencies of five metal kilns, including Siamese, Niligiri, Standard Beehive, South African garage, and commercial half orange from different nations, including Malaya, India, Brazil, South Africa, and Argentina and found variations in charcoal yields that ranged from 12.5 to 30 %, reported as the dry weight ratios of the charcoal output to the wood feedstock input.

4.3.1.3 Retorts

Peterson and Jackson[25] produced biochar by adopting two different processes a retort-type oven with inert gas, and a gasification technology using various crop residues (such as corn stover, wheat straw, and wheat straw treated with glycerin). They reported that gasification is a simpler, easier, and more cost-effective means to produce biochar compared with retort, because while the retort method was effective in the absence of oxygen to promote pyrolysis, it was also costly during scale-up. This expense was due to the need to control the atmosphere with sealed systems in conjunction with the use of inert gas. For operation, as a batch reactor, atmospheric control is also required, otherwise it cannot run with a continuous supply of feedstock. Adam[26] built an improved charcoal production system in India and East Africa from a low-cost retort kiln that produces charcoal from a forest residue like wood and that is eco-friendlier. Also, during the experiment Adam[26] realized that charcoal production efficiency was approximately doubled (30 to 42%) over traditional charcoal production methods (10 to 22%). He also analyzed that ICPS reduces emission into the atmosphere up to 75% as compared to traditional carbonization processes. Antal and Mok[27] studied the required operating cycle for the production of charcoal from the Missouri kiln in the USA and reported that it produces charcoal in a 25 % yield for every 7-12 days during its operating cycle. The average temperature required for the operation is between 450-510 °C; the working temperature varies significantly throughout the kiln, which affects the charcoal quality. Further, Moreira[28] produced biochar from the cashew nut shell in a batch type reactor. The temperature was varied from 200-400 °C and yields 30% biochar, 40% liquid, and 30% gas products.

4.3.2 Continuous Process for Production of Biochar

At present, the continuous process for production of biochar is widely adopted in the commercial sectors due to maximum yield, energy efficiency, and its quality. The biochar yield found between 25 to 35%. As a major benefit the continuous production of biochar is ideal for medium to large-scale production along with a greater flexibility towards biomass feedstock, which are major benefits. Continuous process for biochar production includes,

(a) Drum type pyrolyzer

(b) Screw type pyrolyzer

(c) Rotary kiln

4.3.2.1 Drum Type Pyrolyzer

Robert[29] used a generalized model in which the feedstock is pyrolyzed in continuous operation and is horizontally mounted to a drum kiln and heated externally to around 450 ° C. The continuous feeding and moving of biomass took place in the drum with the help of paddles, which increased kiln efficiency about 50 % so that 90 % of the heat recovered from the kiln was used for drying of the feedstock. Jelinek[30] developed a drum pyrolyzer, which uses heating tubes placed in the center of the durm. The tubes are subjected to low-temperature carbonization of trash and they reuse material with a temperature about 400-500 ° C by slowly rotating the drum. In the case of the drum pyrolyzer feed material to be carbonized was located near one end of the face and discharge took place at the other end to the face. Collin[31] discovered that aromatic pyrolysis oil can be produced by pyrolyzing special wastes containing hydrocarbons such as scrap tires, cable, waste plastics, etc. in an indirectly heated drum reactor at a temperature of around 700 °C. Collin[31] saw a yield of up to 50 % in relation to the organic material. Becchetti[32] studied the use of a conventional type rotary drum pyrolysis reactor for the production of pyrolysis gases and carbonaceous solid residue like charcoal from municipal solid waste and observed that the pyrolysis process not only improved the energy yield but also minimized the waste disposal problem; solid waste was controlled to 10-15% of the total weight o the initial residue.

4.3.2.2 Screw Type Pyrolyzer

Agirre[33] developed an auger reactor for the continuous carbonization process by using biomass waste. During the experiment it was realized that a 900 °C temperature was required for suitable quality of charcoal production, which contains a high carbon content of approximately 85 % and a low volatile amount of approximately 10 %. In the case of auger pyrolysis reactor, there are many parameters such as moisture content, residence time, grain size and operating temperature effects on the yield of charcoal and its quality. Maschio[34] studied a moving-bed in the pilot-plant, a continuous screw reactor for the charcoal production by using biomass with indirect heating. They analyzed that a 350 °C to a 450 °C operating temperature was required for charcoal production. During the study they realized that a heating rate was required in the range of 20-40

K/min; the higher temperature could decrease the charcoal yield and particle size should be range from 50 to 200 mm. Brown and Brown[35] developed a laboratory scale reactor to pyrolyze red oak wood biomass for the production of char and bio-oil. During the experiment, they found that operating conditions like flow rate of sweep gas (3.5 standard L/min), heat carrier temperature (600 °C), high auger speeds (63 RPM), and high heat carrier mass flow rates (18 kg/h) were helpful for maximum bio-oil yield and minimum char yield. The result indicated that this reactor was well suited for bio-oil production which achieved more than 73 % liquid yield. Mozammel[36] used a Herbold pyrolyzer in which screw type shaft was fitted inside the reactor to produce an activated charcoal from feedstock as a coconut shell, using $ZnCl_2$ activation. While performing the experiment, results were obtained in which initial calorific value of coconut shell was 18.38 MJ/kg, and final calorific value of charcoal found was 30.75 MJ/kg. Fixed carbon content was approximately 76.32 % and had a maximum yield up to 32.96 %. The activation time required 50 minutes for the production of activated charcoal at a temperature of 600 °C with an impregnation ratio of about 40 %. More recently, Ferreria[37] developed a screw reactor to produced biochar from elephant grass. The reactor temperature during the experimental study was ranging between 400 °C to 600 °C. Their experimental results reveal that maximum biochar yield was found about 37.4 % at 400 °C.

4.3.2.3 Rotary Kiln

Ortiz[38] carried out a study using the pilot rotary kiln to produce carbonized material from a variety of raw materials such as eucalyptus wood. The pilot rotary kiln was cylindrical and rotated around its longitudinal axis. To facilited the discharge of material, the pilot rotary kiln was slightly inclined (slope about 2-6%). In their research project, Ogawa[39] introduced an internal heating rotary kiln designed to produce charcoal using wood waste as a feedstock. During their experiment, they found that the rotary kiln produced biochar of around 358.0 Mg-C/year from 936.0 Mg-C/year of wood waste at a planned temperature in the range of 500-600 °C. Schimmelpfennig and Glaser[40] analyzed two different rotary kilns used in the carbonization of organic material that discharged pyrolysis gases suitable for heating purposes or for driving the processes. The experiment used rotary kiln that are heated externally and have a shape similar to a cylindrical pyrolyzer in which biomass is moved continuously by rotating the spiral inside the kiln. The rotary kilns produced a total of 16 samples of biochar. Ten samples were produced in a vertically constructed rotary kiln in china oprating at a temperature range from 400°C to 600°C and using bamboo as a feedstock. Another six samples Switzerland were produced from a horizontally constructed kiln heated to a temperature of 650°C.

4.3.3 Novel Processes

This novel process is called flash carbonization. In it, biomass is quickly and efficiently converted into biochar. For this maximum biochar yield was around 40–50% with 70–80% fixed carbon content[41]. Antal[27] examined the high yield of charcoal through the use of different feedstocks, such as Leucaena wood, oak wood, corncob, and macadamia nut shells, carbonized at high pressure (1 MPa) in controlled flash fire, within a packed bed. In flash carbonization, the direction of the fire and the entry of the air was done with a counter current and at an elevated pressure. Charcoal with fixed carbon yield was reached at less than 30 min of reaction time. Furthermore, during the experiment, the yield of charcoal was between 29.5% and 40%, fixed carbon ranged from 27.7% to 30.9%, and the energy conversion efficiency of biomass to charcoal ranged from 55.1% to 66.3%. Wade[42] investigated laboratory-scale flash carbonization (novel process) for the conversion of feedstock biomass (corncob, and macadamia nut shell) into biocarbon. During the experiment, biomass feedstock was placed in a packed bed within a pressure vessel, and an initial pressure of 1 - 2 MPa was maintained through the use of compressed air; flash fire was ignited at the bottom of the bed, and after a duration of 2 minutes, air was supplied to the top of the bed. it was found that biomass could be converted to biocarbon at a high yield. For corncob, a pressure of more than 1.31 MPa was achieved, at a rate up to 1.21 MPa/s, for an initial pressure of the system of 2.17 MPa. In the case of macadamia nut shell this, phenomenon did not occur.

4.4 Method of Biomass Heating to Produce Biochar

The biochar production from crop residues starts with the feeding of biomass into biochar production unit and combustion in the absence of air. The formation of charcoal is completed in five different temperature stages. Stage-1: At 20 to 110 °C, biomass absorbs heat as it dries, giving off moisture as water vapour. At this stage, the temperature remains at or slightly above 100°C until the wood is dry. Stage-2: At 110 - 270 °C, biomass starts to decompose by giving off carbon monoxide, carbon dioxide, acetic acid and methanol, making an endothermic reaction. Stage-3: At 270 – 290 °C, this is the point when an exothermic reaction starts, generating a considerable amount of heat. Such a reaction leads to a continuous breakdown; the desired temperature is maintained to avoid the wood from cooling down below the decomposition temperature. During the exothermic reaction, gases in vapour form are released with some tar. Stage-4: With increasing temperature, vapour mixture of combustible gases (i.e., carbon monoxide, hydrogen and methane) and carbon dioxide are released into the atmosphere. As temperature increases up to 400°C, the condensable vapours such as water, acetic acid, methanol, acetone, etc. and tar are predominate. Stage-5: when the temperature reaches 400°C, the transformation of biomass to charcoal will be practically complete, but appreciable amounts of tar are contained within the biochar, and some tar condensed on charcoal. To avoid this, the temperature should be further increased to 500°C to complete the carbonization stage.

There are several of ways to provide heat to maintain the desired temperature of pyrolysis kilns. One method involves combusting part of the biomass within the kiln. This is called autothermal pyrolysis, as illustrated in Fig.4.4. Due to the use of partial combustion, authothermal kilns typically have lower char yields. Another method is for the heat to be produced externally and to directly heat the biomass. This involves, hot gas being brought into contact with the biomass, as shown in Fig.4.5, or heat being transferred through the reactor walls, as in shown Fig. 4.6. Condensable pyrolysis vapour can be recovered during indirect heating, and this ultimately enhances the biochar yield[43].

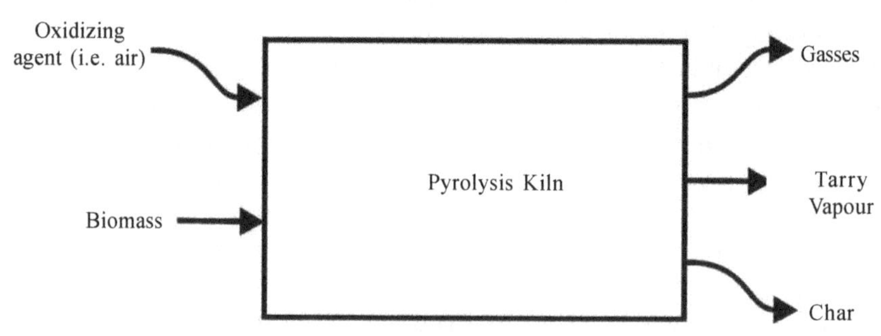

Fig. 4.4: Kiln, Autothermal Carbonization [43]

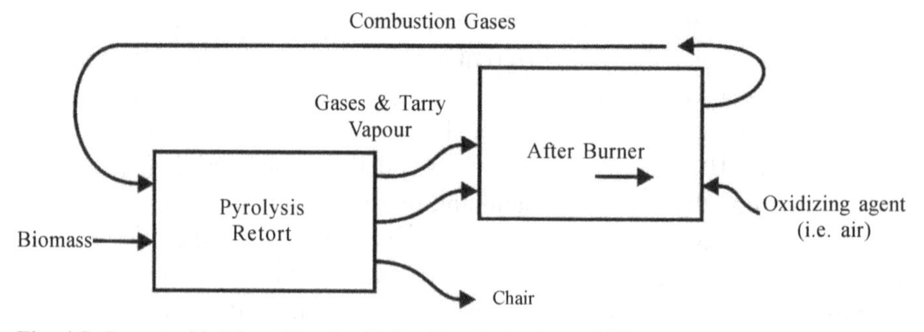

Fig. 4.5: Retort with Direct Heating Using Pyrolysis Gases [43]

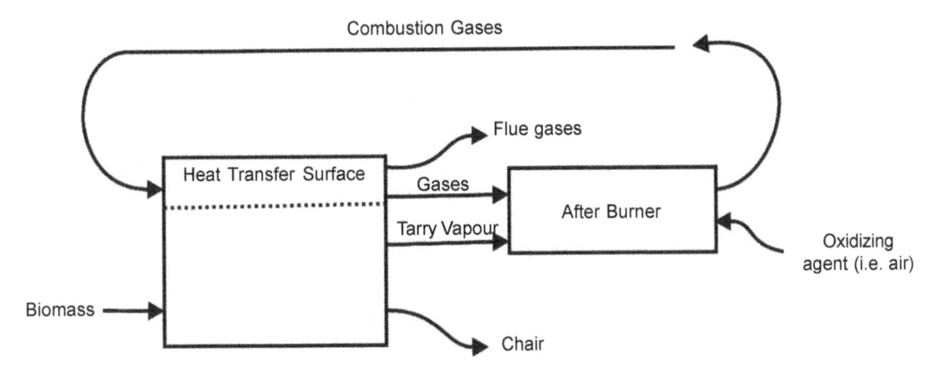

Fig. 4.6: Retort with Indirect Heating Using Pyrolysis Gases [43]

A performance test of the Argentinean-type charcoal kiln (see Fig. 4.7), involving the carbonisation of wood biomass was carried out by Mohod and Panwar[44]. The kiln was tested with Anjan (*Hardwickia binata*), Babul (*Acacia nilotica*), Behada (*Terminalia chebula*), Char (*Buchnania lazan*), and Dhawda (*Anogeissus latifolia*) wood and the mass conversion efficiency was found to be 27.14%.

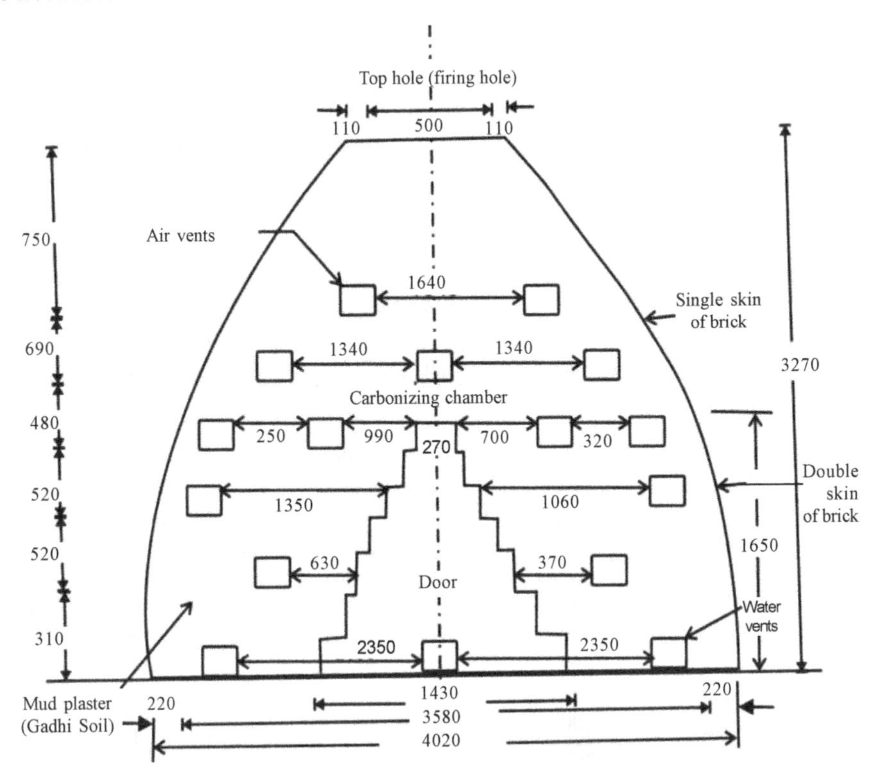

Fig. 4.7: Argentinean type Charcoal Kiln (all dimensions in mm)

4.5. Factors Affecting Biochar Production

The performance results of biochar production which occur via different production technologies broadly depends on the various types of feedstock used, the moisture content of said feedstock, and the operating temperatures and pressure points at which experiments were conducted. Biomass has three main groups, for example: cellulose, hemicellulose, and lignin with trace amounts of extractive and minerals. These propositions are varied depending on the feedstock, a variation which highly affects the biochar yield. Moisture content is another factor which affects the biochar properties and char yield. The moisture content affects the char reaction and is extensively used to produce activated carbon. In fast pyrolysis processes, around 10% moisture content is fairly desirable during the charcoal making process[45]

Production of biochar is a thermochemical process and temperature plays a major role in the properties of biochar and its suitability for soil health. A lab scale study on pyrolysis ability to produce biochar from pin, mixed larch and spruce chips, and softwood pellets was conducted by Masek[46] with temperatures between 350 °C and 550 °C and reported that the stability of biochar increases as temperature increases, and the yield of biochar is independent of temperature. Angin and Sensoz[47] also reported that the chemical and surface properties of biochar are affected by pyrolysis temperature. As the pyrolysis temperature is increased from 400°C to 700 °C, the volatile matter, hydrogen, and the oxygen contents of the biochar were decreased, but the value of fixed carbon was increased. Biomass can't be converted into biochar at low pyrolysis temperature (300 °C) because at this temperature desired carbon frame structure does not developed[48].

Reactor operating temperature play vital role in deciding of fixed carbon and oxygen content of bio-char. It has been found that higher operating temperature have higher fixed carbon content and lower oxygen content as presented in Table 4.1.

Table. 4.1: Effect of Temperature on Biochar Composition

Feedstock	Pyrolysis temperature (°C)	Biochar Elements (%)			
		C	H	N	O
Corn cob[49]	400	75.23	3.37	0.82	14.11
	450	77.84	2.95	0.86	11.45
	500	80.85	2.5	0.97	8.87
	550	82.62	2.25	0.84	7.43
Rapeseed[50]	400	57.95	3.43	5.43	33.16
	450	59.77	2.36	5.12	32.75

(Contd.)

	500	61.98	1.92	4.12	31.78
	550	67.29	1.75	4.35	26.21
Safflower seed[47]	400	68.76	4.07	3.77	23.49
	450	70.43	3.49	3.69	22.39
	500	71.37	2.96	3.91	21.76
	550	72.96	2.67	3.74	20.63
	600	73.72	2.34	3.84	20.10
Conocrpus waste[51]	200	64.20	3.96	0.69	26.60
	400	76.80	2.83	0.87	14.20
	600	82.90	1.28	0.71	6.60
	800	85.00	0.62	0.90	4.90
Wheat straw[52]	400	57.80	3.20	1.50	21.60
	500	70.30	2.90	1.40	17.70
	600	73.40	2.10	1.40	14.90
	700	73.90	1.30	1.20	14.60
Corn straw[52]	400	56.10	4.30	2.40	22.00
	500	58.00	2.70	2.30	21.50
	600	58.60	2.00	2.00	18.70
	700	59.50	1.50	1.60	16.60
Peanut shell[52]	400	58.40	3.50	1.80	21.00
	500	64.50	2.80	1.70	18.50
	600	71.90	2.00	1.60	15.00
	700	74.40	1.40	1.40	14.20

Operating pressure also affects the biochar yield. As increasing both absolute pressure and peak temperature results in a decreased in biochar yield; however, fixed carbon yield increases. Pyrolysis reactor operated under high pressure and high temperature maximizes pyrolysis gas production, but reduces the char yield. Pyrolysis process carried out at high temperature (750 °C) electrical conductivity significantly increased but there is scope vitalization of heavy metal (Zn) with the low melting point[53].

4.6. Energy required for the Production Biochar

Energy consumption, during biochar production obtained from the pyrolysis plant, is the major issue for carbon free emission in the carbon industry. The enthalpy for pyrolysis is the energy required to produce the biochar and syngases that, is mainly depended on biomass and its operating condition. The term enthalpy for pyrolysis or carbonization is the heat or energy used to produce pyrolysis. Doren and Robert[54] revealed that the enthalpy required for the thermal decomposition of oat shell and pine were approximately 1.04 ± 0.18 MJ/kg and 1.61 ± 0.26 MJ/kg respectively However, corresponding energy required for the thermal decomposition were approximately 0.8±0.2 MJ/kg and 1.6±0.3 MJ/kg, respectively as presented in Table 4.2. Laird[55] reported that the net amount of energy required for the pyrolysis process is nearly 15 % of the total energy obtained from dry biomass. Furthermore, Fing[56] estimated that the energy demand for obtaining the biochar from biomass was varied in the range of 1.1-16 MJ/kg, whereas, 44-170 MJ/kg energy was required to produce activated carbon, which was approximately 10 times, higher than that of bio-char production.

Kyoung[57] used swine solid and blended swaine solids (29% rye-grass + 71% swine solids) to produce high temperature value added biochar feedstocks. In the same study[58], the authors demonstrated the energy balance of their study for drying and pyrolysis, revealing that 12.5MJ/kg energy was required for swine solids and 0.5 MJ/kg for blended material to obtain desired value added biochar as mentioned in Table 4.2.

Table 4.2: Energy Required to Produce Biochar from Different Feedstocks.

S.No	Type of feedstock	Mode of process	Energy demand (MJ/kg)	Output products
1.	Oat shell[54]	Pyrolysis	0.8±0.2	Bio-char, oil, gases
2.	Pine[54]	Pyrolysis	1.6±0.3	Bio-char, oil, gases
3.	Swin solid[59]	Carbonization (High temperature)	12.5	Oil. bio-char, gases
4.	Blended swin solid[57] (29 % rye grass + 71 % swin solid)	Carbonization (High temperature)	0.5	Oil. bio-char, gases
5.	Biomass[56]	Carbonization	1.1-16	Bio-char

4.7. Classification of Biochar

The International Biochar Initiative (IBI) broadly classifies biochar based on Carbon storage value, Fertilizer value (P, K, S, and Mg only), Liming value, and Particle size distribution. Further, Anon[60,61] proposed three general classes of biochar on the basis of organic carbon content. In class I type biochar the C_{org} mass fraction is about e”60%, in class II it would be in the range of 30% to < 60% while in class III[62] it would be < 10%.

4.8. Stability of Biochar in Soil

The stability of biochar depends on the conditions of its production and biomass feed stock. Spokas[63] conducted a study on the stability of biochar in soil and found that lower oxygen-to-carbon (O : C) ratio resulted in a more stable biochar material. Conclusively, when the oxygen - to - carbon molar ratio (O : C) is greater than 0.6, biochar will probably possess a half-life on the order of less than 100 years; if the range is 0.2 -0.6, the accepted range of half-life is between 100 and 1000 years. If the molar oxygen-to-carbon ratio is less than 0.2, the half-life will be greater than 1000 years. In this way, the process temperature, i.e. pyrolysis temperature, is highly responsible for biochar stability.

4.9. Application of Biochar

4.9.1 Biochar as Soil Improvement

Biochar improves soil physiology and increases productivity and it also assists with crop residues management. After the application of biochar on soil, many studies report that soil acidity was reduced considerably, and essential mineral uptake increased with residual effects for the following season. In biochar significant quantities of K and minor amount of Mg, Ca, Cu, Zn, and Fe are presented, which showing potential as fertilizer[23].

Stockmann[64] reported that soil contains approximately 2,344 Gt of organic carbon globally and is considered the largest terrestrial pool of organic carbon. Small changes in the soil's organic carbon stock could result in significant impacts on the atmospheric carbon concentration. The sustainability of agricultural production is highly dependent on the physical, chemical and biological integrity of the soil. Organic carbon plays a major role in maintaining these factors. The efficient conversion of surplus crop residues as a source of organic carbon is one way to improve the soil health and retain the water-holding capacity, as well as essential nutrients. Demand for food has drastically increased as the global population has grown. Growers are increasingly using chemical fertilizer in soil to meet demand. Soil fertility has significantly decreased as a result of this. The addition of organic carbon is the only option for overcoming this issue. Biochar has tremendous potential to improve soil health and it is currently attracting considerable interest globally due to its sustainable stability of carbon, which also helps in reducing atmospheric carbon dioxide concentration. In the present context, the biochar is considered as a soil amendment tool globally because it has a suitable cation exchange capacity, which improves soil pH, the water holding capacity, and affinity for micro- and macro-plant nutrients[65].

Biochar worked to increase available water, build soil organic matter, enhance nutrient cycling, lower bulk density, act as a liming agent, and reduce leaching of nutrients to ground water. The application of biochar as a soil amendment significantly increased crop yield, even in the absence of nitrogen fertilizer. Mogami[66] reported that the soil water retention capacity with palm shell biochar application is significantly higher than that of without biochar. Further, it was also found that biochar application at 0-10 cm greatly controlled leaching of mobile nutrients such as potassium, thus improving water use efficiency, nutrient availability and plant growth. Jia[67] conducted a pot experiment to estimate the effect of maize straw biochar application on the nitrous oxide (N_2O) and methane (CH_4) emission, N_2O emission factor and vegetable yield. They concluded that biochar application greatly reduced N_2O emissions and N_2O—N emission factors while maintaining vegetable production. They found out that the methane emission was not affected by biochar amendment.

The environmental benefits can be maximized, if recycling of organic wastes occurs through proper routes. Maraseni[68] gives a very good example of value addition in the Australian pulp and paper industry. Promoting fast growing species on plantations can not only aid in achieving goals for greenhouse gas mitigation, but also help in carbon sequestration. Such a value addition approach increases employment in rural areas.

Nitrogen (N) is one of the most important elements that play a major role in plant growth and productivity, as plants use up inorganic N through the root system directly. Nguyen[69] reviewed the effects of biochar on soil inorganic nitrogen and found that biochar production temperature and biochar surface properties are main factors affecting soil inorganic nitrogen. So far, there are limited long-term studies of greater than one year, available in literature, thus the long-term effects of biochar on soil inorganic nitrogen still remain unclear.

Shanta[70] demonstrated the effects of biochar, along with plant growth promoting rhizobacterial, on plant growth variables (i.e. height, stand count, dry biomass). The 20 Mg ha^{-1} biochar treatment, in combination with 100 kg N fertilizer ha^{-1}, found almost the same biomass yield as the treatment with 50 kg N fertilizer ha^{-1} without biochar. Furthermore, it was reported that this effect was not consistent across study sites, highlighting the incomplete understanding of crop responses to biochar application at different study locations. Surprisingly, inoculation of switchgrass seeds with bio-fertilizers did not appear to improve crop yield in the presence or absence of biochar soil amendments.

4.9.2 Application of Biochar in Water Treatment

Recently, biochar derived from biomass has been given significant attention, especially for the effective removal of heavy metals, toxic elements, and contaminants from water and waste water. Biochar is a promising low cost and effective material with remarkable physiochemical properties such as high surface area, cation exchange capacity, aromatic character, carbon content and low H/C ratio, etc.

Wood based biochar as an emerging bio sorbent which has a potential to remove toxic elements from water and waste water. The biochar material's high surface area and its reactivity further its uses in water filters for the removal of pathogens like lipids, and phenol from water. Werner[71] carried out field experiment in Ghana using biochar filtered water for irrigation, and measured the increase in maximum crop yield (more than 40 %) in leafy green vegetable production. Gwenzi[72] asserted that biomass derived biochar based water treatment systems are a potentially low cost sustainable technology for the provision of clean water. Lee[73] conducted experiments for removal of natural organic matters in water through biochar with different doses and reported that at 200/ mg-C/L, biochar removes 90% of organic matter in 20/ min of contact, with a capacity of 0.0064/ mg-dissolve organic carbon /mg-C. Biochar is acted as a super sorbent with the ability to remove organic and inorganic contaminants from the soil as well as water due to its physiochemical properties.

4.9.3 Biochar for Climate Change Mitigation

To avoid the worst consequences of climate changes, human need to significantly reduce global warming emissions and if possible, remove existing carbon dioxide from the atmosphere. Scientist have discovered a more environmentally friendly way to create charcoal by heating biomass, plants, and animal manure in a low-oxygen environment. Biochar amendment on forest floors in an acidic spruce ecosystem could lead to an increase of surface carbon stocks[73]. It is a well-known fact that airborne black carbon, or soot, is a significant contributor to global warming. If biochar is simply spread on top of soil, there is the possibility for airborne black carbon. However, such issues can be avoided, if biochar is tilled deep into the soil, which can also improve the soil's water retention and reduce leaching of agricultural nutrients[74]. Further, Waters[75] reported that issues of climate change mitigation impacts arise largely with the stabilization of soil organic matter using biochar, and generating renewable fuels, which can displace fossil fuel consumption.

4.9.4 Carbon Sequestration

Climate change is one of the biggest challenges presently. It affects entire cropping patterns across the globe. Biomass is usually considered to be a carbon-neutral material; whereas biochar, which is produced through crop residues with stable carbon and returned to soil will act as a long-term sink for atmospheric carbon dioxide. It will enhance carbon fixation and reduce the emission of gases such as CH_4, N_2O, and CO_2. Lehmann[75] reported that the global carbon sequestration potential using agricultural and forestry wastes was estimated at about 0.16 Gt on an annual basis. Further, these authors also reported that using renewable fuels, by the year 2100, the carbon sequestration potential may reach in the range of 5.5 to 9.5 Gt yr^{-1}. Smith[76] estimated the carbon sequestration potential considering agricultural soils globally at about 1.4 to 2.9 Gt of CO_2 equivalents. Chatterjee[77] suggested a sequestration potential of agricultural soils of up to 6 Gt of CO_2 equivalents per year by 2030. An experimental investigation of carbon sequestration through silage maize was carried out under Denmark's climatic conditions by Kristiansen[78]. They found that carbon from maize roots and stubble accumulated in the soil at a rate of 0.25 - 0.49 t C ha^{-1} yr^{-1}. Further, with the addition of 8 t of dry matter per hectare, the carbon accumulation rate was found to be between 0.71 and 0.98 t C ha^{-1} yr^{-1}. Boddey[79] conducted experiments in a subtropical region of Southern Brazil to assess the soil's organic carbon potential at 30 and 100 cm. Soil carbon accumulation rate at a depth of 30 cm was estimated to be between 0.04 and 0.88 mg ha^{-1} yr^{-1} whereas, at 100 cm depth it was between 0.48 - 1.53 mg ha^{-1} yr^{-1}.

4.9.5 Mitigation of Greenhouse Gas Emissions

Carbon dioxide, methane and nitrous oxide are considered to be major greenhouse gases, which are primarily responsible for climate change. The concentration of greenhouse gases in the atmosphere has reached an alarming level. The atmospheric carbon dioxide concentration has increased from 280 ppm prior to industrialization to 379 ppm in 2005 and 402.9 ppm in 2016. Carbon dioxide levels today are higher than at any other point in at least the past 800,000 years[80]. Global crop residues produce about 3758 million tons of carbon dioxide a year, which is equivalent to what is produced by approximately 7560 million barrels of oil. The energy equivalent of these yearly crop residues was estimated to be about 69.9 EJ.

On the Asian continent, straw burning is a common problem. Gupta[81] reported the particulate matter produced by burning one ton of straw 60 kg CO, 1,460 kg CO_2, 199 kg ash and 2 kg SO_2. Furthermore, Gadde[82] estimated that the burning of crop residues on the Asian continent annualy contributes about 0.10 Tg of SO_2, 0.96 Tg of NO_2, 379 Tg of CO_2, 23 Tg of CO and 0.68 Tg of CH_4. Emission of such gases and aerosols adversely affects regional environments and is also responsible for global climate change. Renewable energy harvesting of surplus crop residues, forest residues, and agro industrial wastes has been encouraged in order reduce greenhouse gases.

In 2009, Roberts[29] conducted a life-cycle assessment on the application of biochar derived from stover, switchgrass and yard waste. They reported that it has a much lower greenhouse gas abatement costs of about EUR 30, 45 and 1.5 per tonne CO_2e for biochar derived from stover, switchgrass and yard waste, respectively. Furthermore, Cowie[83] examined the greenhouse gas mitigation potential of poultry litter biochar applied to maize crop and reported a reduction of 3.2 kg CO_2e per kg of biochar.

Zhang[84] conducted an experiment to assess the effect of biochar with and without application of nitrogen on net greenhouse gas balance and greenhouse gas intensity under Jerusalem Artichoke Bioenergy Cropping System. During their experiment, it was found that soil CH_4 emissions were 72 to 80 percent lower in the biochar amended plots than the unamended plots. Furthermore, it was reported that biochar amended soil improved greenhouse gas sink capacity.

4.10. Safety Measures During Biochar Production and Its Applications

There is limited literatures available with details on the smooth operation and necessary safety measures for the production of biochar. Many reports have shown that the moisture content of crop residue or feedstock should be less than 8 % As feedstock with lower moisture content requires less energy to convert into bio-char. It is well known that biochar is produced by heating of biomass, and a considerable amount of smoke is generated during the process. Therefore, the work place should be well ventilated. Sigmund[85] investigated the cytotoxicity of biochar and reported that cytotoxic effects were likely related to its particulate nature and size distribution. They also suggested that, to minimize the risk of exposure, operators should wear respiratory protective equipment during biochar production and its application in the field. It was also suggested that biochar should be applied as a slurry and properly mixed with a soil matrix to avoid secondary dust formation.

References

[1] Harmful Effects of Chemical Fertilizer. (Available from: http://www.bkpindia.net/Pdf/Effect_of_chemical_fertilizer.pdf)

[2] Ji C, Cheng K, Nayak D, Pan G. Environmental and economic assessment of crop residue competitive utilization for biochar, briquette fuel and combined heat and power generation. Journal of Cleaner Production 2018; 192: 916-23.

[3] Hiloidhari M, Das D, Baruah DC. Bioenergy potential from crop residue biomass in India. Renewable and Sustainable Energy Reviews 2014; 32: 504-12.

[4] Neves EG, Bartone RN, Petersen JB, Heckenberger MJ. The timing of Terra Preta formation in the central Amazon: new data from three sites in the central Amazon. Springer: Berlin; London 2004.

[5] Woolf D, Amonette JE, Street-Perrott FA, Lehmann J, Joseph S. Sustainable biochar to mitigate global climate change. Nat. Commun 2010: 1:56.

[6] Klark M, Rule A. The technology of wood distillation. London Chapman & Hall Ltd 1925.

[7] Boroson ML, Howard JB, Longwell JP, Peters WA. Heterogeneous cracking of wood pyrolysis tar over wood char surface. Energy & Fuels 1989; 3: 735-740.

[8] Prins MJ, Ptasinski KJ, Janssen FJJG. Torrefaction of wood: Part1. Weightloss kinetics. J Anal Appl Pyrolysis 2006; 77: 28–34.

[9] Onay O, Kockar OM. Slow, fast and flash pyrolysis of rapeseed. Renew Energy 2003; 28: 2417–33.

[10] Lehmann J, Stephen J. Biochar for Environmental Management: An Introduction. Biochar for Environmental Management-Science and Technology, UK, Earthscan 2009.

[11] Woolf D, Amonette JE, Street-Perrott FA, Lehmann J, Joseph S. Sustainable biochar to mitigate global climate change. Nat. Commun 2010; 1:56.

[12] Steiner C, Bellwood-Howard I, Häring V, Tonkudor K, Addai F, Atiah K, Abubakari AH, Kranjac-Berisavljevic G, Marschner B, Buerkert A. Participatory trials of on-farm biochar production and use in Tamale, Ghana. Agronomy for Sustainable Development 2018; 38(1): 12.

[13] Phillips CL, Trippe K, Reardon C, Mellbye B, Griffith SM, Banowetz GM, Gady D. Physical feasibility of biochar production and utilization at a farm-scale: A case-study in non-irrigated seed production. Biomass and Bioenergy 2018; 108: 244-51.

[14] Basu P. Biomass gasification, pyrolysis and torrefaction: practical design and theory. Academic Press 2013.

[15] Kammen DM, Lew DJ. Review of Technologies for the Production and Use of Charcoal. Renewable and appropriate energy laboratory report, (http://rael.berkeley.edu/old_drupal/sites/default/files/old-site-files/2005/Kammen-Lew-Charcoal-2005.pdf assessed on December 15, 2017)

[16] Duku MH, Gu S, Ben HE. Biochar production potential in Ghana: a review. Renew. Sust. Energy Rev. 2011;15: 3539–3551.

[17] Masek O. Biochar production technologies, http://www.geos.ed.ac.uk/sccs/ biochar/documents/BiocharLaunch-OMasek.pdf. (assessed on December 15, 2017)

[18] Bailis R. Modeling climate change mitigation from alternative methods of charcoal production in Kenya. Biomass and Bioenergy 2009; 33(11): 1491-1502.

[19] Lohri CR, Rajabu HM, Sweeney DJ, Zurbrügg C. Char fuel production in developing countries–A review of urban biowaste carbonization. Renewable and Sustainable Energy Reviews 2016; 59: 1514-1530.

[20] FAO (United Nations Food and Agriculture Organization): Simple Technologies for charcoal making 1983; 41.(www.fao.org/docrep/S5328e/x5328e00.htm, accessed on December 16, 2017)

[21] Kristoferson LA, Bokalders V. Renewable energy technologies: their applications in developing countries. Elsevier 2013.

[22] Deal C, Brewer CE, Brown RC, Okure MA, Amoding A. Comparison of kiln-derived and gasifier-derived biochars as soil amendments in the humid tropics. Biomass and Bioenergy 2012; 37: 161-168.

[23] Pennise DM, Smith KR, Kithinji JP, Rezende ME, Raad TJ, Zhang J, Fan C. Emissions of greenhouse gases and other airborne pollutants from charcoal making in Kenya and Brazil. Journal of Geophysical Research: Atmospheres 2001; 106(D20): 24143-24155.

[24] Mwampamba TH, Owen M, Pigaht M. Opportunities, challenges and way forward for the charcoal briquette industry in Sub-Saharan Africa. Energy for Sustainable Development 2013;17(2): 158-170.

[25] Peterson SC, Jackson MA. Simplifying pyrolysis: Using gasification to produce corn stover and wheat straw biochar for sorptive and horticultural media. Industrial Crops and Products 2014; 53: 228-235.

[26] Adam JC. Improved and more environmentally friendly charcoal production system using a low-cost retort–kiln (Eco-charcoal). Renewable Energy 2009; 34(8): 1923-1925.

[27] Antal MJ, Grønli M. The art, science, and technology of charcoal production. Industrial & Engineering Chemistry Research 2003; 42(8): 1619-1640.

[28] Moreira R, dos Reis Orsini R, Vaz JM, Penteado JC, Spinacé EV. Production of biochar, bio-oil and synthesis gas from cashew nut shell by slow pyrolysis. Waste and Biomass Valorization 2017; 8(1): 217-224

[29] Roberts KG, Gloy BA, Joseph S, Scott NR, Lehmann J. Life cycle assessment of biochar systems: estimating the energetic, economic, and climate change potential. Environmental Science and Technology 2009; 44: 827-833.

[30] Jelinek H (1989) Pyrolysis system.U.S. Patent No. 4,840,129. 20 Jun. 1989. (https://www.google.com/patents/US4840129 , accessed on December 20, 2017)

[31] Collin G. Pyrolytic recovery of raw materials from special wastes.1980; 479-484.

[32] Becchetti F, Von Christen FE. Integrated process for waste treatment by pyrolysis and related plant. U.S. Patent No. 7,878,131. 1 Feb. 2011.(https://patents.google.com/patent/US20090020052A1/en, accessed on December 18, 2017)

[33] Agirre I, Griessacher T, Rösler G, Antrekowitsch J. Production of charcoal as an alternative reducing agent from agricultural residues using a semi-continuous semi-pilot scale pyrolysis screw reactor. Fuel processing technology 2013; 106: 114-121.

[34] Maschio G, Koufopanos C, Lucchesi A. Pyrolysis, a promising route for biomass utilization. Bioresource technology 1992; 42(3): 219-231.
[35] Brown JN, Brown RC. Process optimization of an auger pyrolyzer with heat carrier using response surface methodology. Bioresource technology 2012; 103(1): 405-414.
[36] Mozammel HM, Masahiro O, Bhattacharya SC. Activated charcoal from coconut shell using ZnCl 2 activation. Biomass and Bioenergy 2002; 22(5): 397-400.
[37] Ferreira SD, Manera C, Silvestre WP, Pauletti GF, Altafini CR, Godinho M. Use of Biochar Produced from Elephant Grass by Pyrolysis in a Screw Reactor as a Soil Amendment. Waste and Biomass Valorization 2008:1-2 doi.org/10.1007/s12649-018-0347-1
[38] Ortiz OA, Suárez GI, Nelson A. Dynamic simulation of a pilot rotary kiln for charcoal activation. Computers & chemical engineering 2005; 29(8): 1837-1848.
[39] Ogawa M, Okimori Y, Takahashi F. Carbon sequestration by carbonization of biomass and forestation: three case studies. Mitigation and adaptation strategies for global change 2006; 11(2): 421-436.
[40] Schimmelpfennig S, Glaser B. One step forward toward characterization: some important material properties to distinguish biochars. Journal of Environmental Quality 2012; 41(4): 1001-1013.
[41] Evans RJ. The relation of pyrolysis processes to charcoal chemical and physical properties. National Renewable Energy Laboratory, diakses pada 2008: 18.
[42] Wade SR, Nunoura T, Antal MJ. Studies of the flash carbonization process. 2. Violent ignition behavior of pressurized packed beds of biomass: A factorial study. Industrial & engineering chemistry research 2006; 45(10): 3512-3519.
[43] Ronsse F. Report on biochar production techniques. A publication of the Interreg IVB project Biochar: climate saving soil. Ghent University; 2013.
[44] Mohod AG, Panwar NL. Evaluation of traditional half orange type charcoal kiln for carbonisation: a case study. World Review of Science, Technology and Sustainable Development 2011; 8:196–202.
[45] Bridgwater AV, Peacocke GVC. Fast Pyrolysis Processes for Biomass. Renewable and Sustainable Energy Reviews 2000; 4:1-73.
[46] Masek O, Brownsort P, Cross A, Sohi S. Influence of production conditions on the yield and environmental stability of biochar. Fuel 2013; 103:151-5.
[47] Angýn D, Þensoz S. Effect of pyrolysis temperature on chemical and surface properties of biochar of rapeseed (Brassica napus L.). International Journal of phytoremediation 2014; 16(7-8):684-93.
[48] Tan Z, Yuan S. The Effect of Preparing Temperature and Atmosphere on Biochar's Quality for Soil Improving. Waste and Biomass Valorization 2017; 1-11 doi.org/10.1007/s12649-017-0145-1
[49] Zheng W, Sharma BK, Rajagopalan N. Using Biochar as a Soil Amendment for Sustainable Agriculture. Submitted to the Sustainable Agriculture Grant Program Illinois Department of Agriculture 2010.
[50] Angin D. Effect of pyrolysis temperature and heating rate on biochar obtained from pyrolysis of safflower seed press cake. Bioresour Technol 2013; 128: 593–597.
[51] Al-Wabel MI, Al-Omran A, El-Naggar AH, Nadeem M, Usman A. Pyrolysis temperature induced changes in characteristics and chemical composition of biochar produced from conocarpus wastes. Bioresour Technol 2013; 131:374–379.
[52] Gai X, Wang H, Liu J, Zhai L, Liu S, Ren T, Liu H. (2014) Effects of Feedstock and Pyrolysis Temperature on Biochar Adsorption of Ammonium and Nitrate. Plos One 2014; 9(12): e113888. doi:10.1371/journal.pone.0113888
[53] Wang K, Peng N, Lu G, Dang Z. Effects of Pyrolysis Temperature and Holding Time on Physicochemical Properties of Swine-Manure-Derived Biochar. Waste and Biomass Valorization 2020; 11: 613-624.

[54] augaard DE, Brown RC (2003) Enthalpy for pyrolysis for several types of biomass. Energy & Fuels 17(4): 934-939

[55] Laird DA (2008)The charcoal vision: a win–win–win scenario for simultaneously producing bioenergy, permanently sequestering carbon, while improving soil and water quality. Agronomy journal 100(1):178-181

[56] Fing. Biochar vs Activated Carbon 2016. http://fingerlakesbiochar.com/biochar-vs-activated-carbon/ (assessed on Oct 2, 2018)

[57] Ro KS, Cantrell KB, Hunt PG. High-temperature pyrolysis of blended animal manures for producing renewable energy and value-added biochar. Industrial & Engineering Chemistry Research 2010; 49(20): 10125-10131

[58] Cheng X, Tang Y, Wang B, Jiang J. Improvement of charcoal yield and quality by two-step pyrolysis on rice husks. Waste and Biomass Valorization 2018; 9(1): 123-130

[59] Liang B, Lehmann J, Solomon D, Sohi S, Thies JE, Skjemstad JO, Luizão FJ, Engelhard MH, Neves EG, Wirick S. Stability of biomass-derived black carbon in soils. Geochim Cosmochim Acta 2008; 72(24):6069–6078.

[60] Anon. Standardized Product Definition and Product Testing Guidelines for Biochar that Is Used in Soil, IBI-STD-01.1, International Biochar Initiative, Westerville, OH 2003.

[61] Anon. Standardized Product Definition and Product Testing Guidelines for Biochar that Is Used in Soil, IBI-STD-2.0, International Biochar Initiative, Westerville, OH 2014.

[62] Klasson KT. Biochar characterization and a method for estimating biochar quality from proximate analysis results. Biomass and Bioenergy 2017; 96:50-58.

[63] Spokas KA. Review of the stability of biochar in soil: predictability of O:C molar ratios. Carbon management 2010; 1(2): 289-303.

[64] Stockmann U, Adams MA, Crawford JW. The knowns, known unknowns and unknowns of sequestration of soil organic carbon. Agriculture, Ecosystems and Environment 2013; 164: 80– 99

[65] Nsamba HK, Hale SE, Cornelissen G, Bachmann RT. Sustainable Technologies for Small-Scale Biochar Production—A Review. Journal of Sustainable Bioenergy Systems 2015; 5: 10-31.

[66] Mogamia A, Tanoa Y, Matsumotoa H, Nishiharaa E. Improvement of sandy- soil water and nutrient use efficiency using palm shell biochar under controlled moisture conditions. Asia Pacific Biochar Conference, apbc Kyoto 2011.

[67] Jia J, Li B, Chen Z, Xie Z, Xiong Z. Effects of biochar application on vegetable production and emissions of N_2O and CH_4. Soil Science and Plant Nutrition 2012; 58 (4): 503-509.

[68] Maraseni TK. Biochar: maximising the benefits. Int J Environmental Studies 2010; 67(3): 319-327.

[69] Nguyen TTN, Xu CY, Tahmasbian I, Che R, Xu Z, Zhou X, Wallace HM, Bai SH Effects of biochar on soil available inorganic nitrogen: A review and meta-analysis. Geoderma 2017; 288:79–96.

[70] Shanta N, Schwinghamer T, Backer R, Allaire SE, Teshler I, Vanasse A, Whalen J, Baril B, Lange S, MacKay J, Zhou X, Smith DL. Biochar and plant growth promoting rhizobacteria effects on switchgrass (Panicum virgatum cv. Cave-in-Rock) for biomass production in southern Quebec depend on soil type and location. Biomass and Bioenergy 2016; 95:167-173.

[71] Werner S, Kätzl K, Wichern M, Marschner B. Biochar in waste water treatment to produce safe irrigation water, recover nutrients and reduce environmental impacts of waste water irrigation. In EGU General Assembly Conference Abstracts 2018; 20, 15820

[72] Gwenzi W, Chaukura N, Noubactep C, Mukome FN. Biochar-based water treatment systems as a potential low-cost and sustainable technology for clean water provision. Journal of environmental management 2017; 197: 732-749

[73] Bruckman VJ, Terada T, Uzun BB, Apaydýn-Varol E, Liu J. Biochar for climate change mitigation: tracing the in-situ priming effect on a forest site. Energy Procedia 2015; 76: 381-387.
[74] Ernsting A, Smolker R. Biochar for climate change mitigation: fact or fiction. Agrofuels and the Myth of the Marginal Lands 2009: 1-10.
[75] Lehmann J, Gaunt J, Rondon M. Bio-char sequestration in terrestrial ecosystems–a review. Mitigation and adaptation strategies for global change 2006; 11(2): 395-419.
[76] Smith P, Martino D, Cai Z, Gwary D, Janzen H, Kumar P, McCarl B, Ogle S, O'Mara F, Rice C, Scholes B, Sirotenko O. Chapter 8: Agriculture. In: Metz B, Davidson OR, Bosch PR, Dave R, Meyer LA . Climate Change 2007: Mitigation. Contribution of Working Group III to the Fourth Assessment Report of the Intergovernmental Panel on Climate Change. Cambridge, New York 2007.
[77] Chatterjee A, Lal R. On farm assessment of tillage impact on soil carbon and associated soil quality parameters. Soil Tillage Res. 2009; 104: 270–277.
[78] Kristiansen S.M.H.E.M, Jensen LS, Christensen BT. Natural ^{13}C abundance and carbon storage in Danish soils under continuous silage maize. Eur. J. Agron. 2005; 22: 107–117.
[79] Boddey RM, Jantalia CP, ConceiC ao PC, Zanatta JA, Bayer C, Mielniczuk J, Dieckow J, Dos Santos HP, Denardin JE, Aita C, Giacomini SJ, Alves BJR, Urquiaga S. Carbon accumulation at depth in Ferrasols under zero-till subtropical agriculture. Global Change Biol 2010; 16(2)P: 784–795.
[80] Lindsey R. Climate Change: Atmospheric Carbon Dioxide. (https://www.climate.gov/news-features/understanding-climate/climate-change-atmospheric-carbon-dioxide (Retrieve on December 14, 2017)
[81] Gupta PK, Sahai S, Singh N, Dixit CK, Singh DP, Sharma C. Residue burning in rice-wheat cropping system: Causes and implications. Current Science 2004; 87(12): 1713–1715.
[82] Gadde B, Bonnet S, Menke C, Garivait S. Air pollutant emissions from rice straw open field burning in India, Thailand and the Philippines. Environmental Pollution 2009; 157: 1554–1558.
[83] Cowie AL, Cowie AJ, Solutions RC. Life cycle assessment of greenhouse gas mitigation benefits of biochar. Report to IEA Bioenergy Task, 2013; 38: 136-145.
[84] Zhang Y, Lin F, Wang X, Zou J, Liu S. Annual accounting of net greenhouse gas balance response to biochar addition in a coastal saline bioenergy cropping system in China. Soil and Tillage Research 2016; 158: 39-48.
[85] Sigmund G, Huber D, Bucheli TD, Baumann M, Borth N, Guebitz GM, Hofmann T. Cytotoxicity of biochar: a workplace safety concern?. Environmental Science & Technology Letters 2017; 4(9): 362-366.

5

Pyrolysis Process

5.1 Introduction

Today, energy distribution and environmental issues have developed into international threats. Demand for alternative fuels increases every day due to rapid population growth, rising living standards, and economical sustainability. With its massive population, India has become the fourth largest energy consumer country in the world after the United States, China, and Russia. Because most nations use fossil fuels to fulfill their energy demands, they create negative impacts on the environment including raising global temperatures, pollution, acid rain, and other effects. As such, the quest for alternative fuels derived from renewable resources has captured significant attention throughout the world[1]. Biomass as a renewable energy sources is an important asset for achieving economic development is due to the availability of biomass at a low cost, a good conversion efficiency, the addition of jobs, and an increase in the biodiversity[2,3]. Typically, biomass is composed of cellulose (40-50%), hemicelluloses (25-35%), and lignin (15-30%). Large amount of resources for biomass are available such as woody biomass, crop residues and their by-products, food processing waste, municipal solid waste, aquatic plant etc.

In the present context, pyrolysis process has been received a remarkable identity as a potential method for the conversion of any organic biomass into valuable energy rich products due to its simple in operation and also required a reasonable cost for conversion process. Pyrolysis is the thermal decomposition of any organic material at a specific temperature in absence of air/oxygen; the process ends with three resulting products, namely liquid (bio-oil), solid (biochar), and syngases[4]. Recently, biooil produced through pyrolysis process and further it up grading has been attracted a significant attention due to its major use as biofuel and as a precursor material for making chemicals[5]. The pyrolysis process also classified in three different types as per its operating conditions; slow, fast and flash pyrolysis.

The primary end product of the pyrolysis process is liquid oil known as biooil or pyrolytic oil, which has a dark brown colour and potential as an alternative fuel for multiple applications. According to International Energy agency News[6], the consumption of biofuels around the world will increase from the current 2 % of the total global share of fuel to 27 % in 2050. Biooil or pyrolytic oil is composed of organic substances like phenol, amines, ketones, ethers, esters, furans, aromatic hydrocarbons, alcohols, and water. Despite advantages that include being eco-friendly, having a low cost requirement, and having high conversion efficiency, biooil is facing some technical challenges and limitations for use in the commercial sector due to its high water content (15-30 %) and high composition of oxygenated compounds (35-60%) such as acids, aldehydes, ketones, and alcohols. Excess level of some components can result in unfavorable effects on biooil characteristics: such as lower calorific value, which decreased the combustion efficiency: and instability, which is sometimes due to oxygenated compound corrosion occurs[7].

5.2 Current Technical Status of Biomass in India

In India, people are consuming nearly 32 % primary energy obtained from biomass and almost 70 % of people in the country are depending upon the energy derived from biomass. In developing countries like India, biomass has a huge potential and considered as inexpensive, carbon-neutral, and abundant source of energy. Mainly available biomass material from agriculture waste includes rice husk, bagasse, straw, soya husk, groundnut shell, sawdust, coffee waste etc. However, according to Ministry of New and Renewable Energy (MNRE)[8], the current potential production of biomass in India is about 500 million metric tonnes per annum; among this only small fraction of biomass has been further utilized as an animal feedstuff, in small scale industrial application, and most commonly used as a domestic cooking fuel. Among the total available biomass, most of the biomass is remains unutilized known as surplus biomass nearly ≈ 120-150 MT and however it creates a disposal related problems. This surplus biomass quantity nearly equal to 25 Exajoule (EJ) of the total energy potential, which is about 10 % of India's total primary energy requirement. Otherwise, as a result, a very common solution preferred by some peasants is the burning of surplus residue in open field which significantly creates health issues throughout the living organisms. As per the information from Intergovernmental Panel on Climate Change (IPCC), almost 25 % surplus residue from crop residue is put through for field burning all around the world, which subjected to about 36 EJ of energy is being wasted annually[9]. According to MNRE, the agriculture or forest surplus residue have the potential to generate power about 18,000 MW.

5.3 Classification of Pyrolysis Process

Pyrolysis process can efficiently operate under different operating conditions, so there are different types of pyrolysis process such as slow, intermediate, fast, flash, vacuum, ultra-flash, catalytic pyrolysis etc. Vacuum pyrolysis is carried out at a very low pressure (up to 4 kPa), while atmospheric pyrolysis of biomass is conducted under atmospheric pressure. Waste material including forest waste, woody biomass and agricultural waste are considered as most suitable primary feedstock for the pyrolysis process. A complete biomass pyrolysis refinery is illustrated in Fig. 5.1

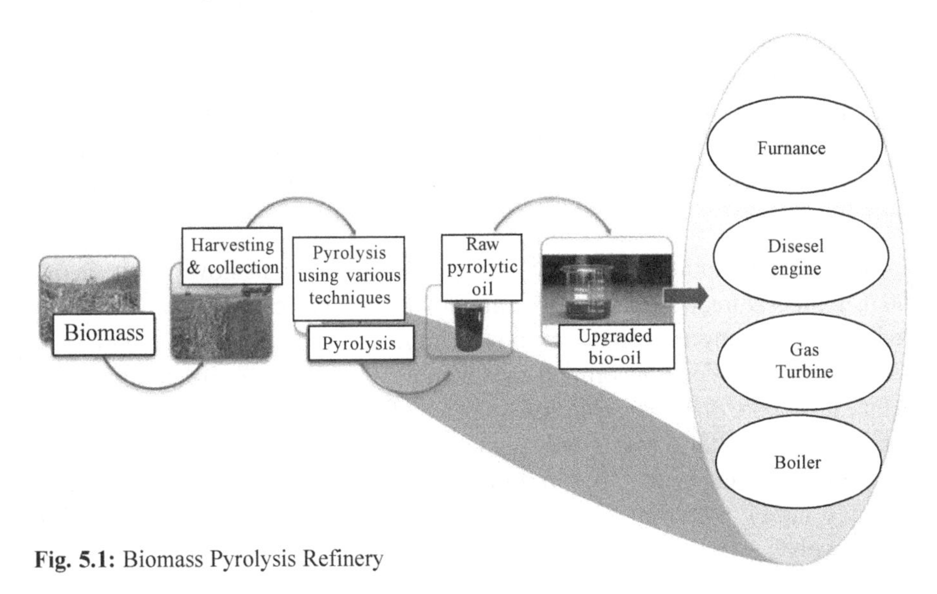

Fig. 5.1: Biomass Pyrolysis Refinery

5.3.1 Slow Pyrolysis

The slow pyrolysis of biomass conveys a number of advantages because it produces a primary end product, biochar, which is an organic carbon-rich product that can act as carbon sequester and improve the soil quality. Slow pyrolysis is performed at a moderate temperature, with a long residence time and a low heating rate; therefore, biochar is the primary end product of this process. The most commonly used reactors for slow pyrolysis are rotary kilns, drums, and screw reactors. However, according to Marshall[10] an auger/screw reactor is considered to be more suitable for both slow and fast pyrolysis processes. Generally, biochar is produced during slow pyrolysis in the absence of air by maintaining the temperature between 300-550 ^{0}C; therefore, the complete reaction requires a long residence time (5-30 min) and a very low heating rate (0.1-0.8 ^{o}C/s)[11]. The yields for biochar and biooil are greatly influenced by the processing temperature, the feedstock properties, the pyrolysis environment (including the availability of inert media), and the residence time. The end products of pyrolysis processes are primarily biochar (35% of the yield), followed by biooil (30% of the yield) and syngas (35% of the yield). Recently, Yu et al.[12] performed slow pyrolysis on hinoki cypress (*chamaecyparis abtusa*), in a fixed bed reactor by keeping temperature range between 350-600 °C and noticed that the yield of biochar decreased from 32.7% (350 °C) to 20.7% (600 °C) as the temperature increased. Simultaneously, the biooil yield increased from 28.8% (350°C) to 35.5% (600 °C). Despites these advantageous results, the produced biooil contained a higher proportion of water. The biochar produced by slow pyrolysis (350-450 °C) fulfills the quality criteria for agricultural soil. In addition, biochar produced by slow pyrolysis can act as a carbon sequester because it contains the maximum amount of organic carbon (> 50%), O/C molar ratio (< 0.4), and H/C molar ratio (<0.7). Mass yield is generally influenced by pyrolysis temperature, as temperature increased, the rate of devolatilization gradually increased.

Similarly, biochar obtained at high pyrolysis temperature in slow pyrolysis (at 600 °C temperature, heating rate 3 °C/min for 1 h residence time) using different feedstock including rice straw, corn cob, bamboo chips, eucalyptus bark etc. exhibited good pesticide adsorption capacity due to its properties like aromaticity, surface area, pore diameter, polarity, weak acid fraction, pH etc. Further, authors carried out elemental analysis and characterization of biochar in terms of pH(8-10), cation exchange capacity (36-74 $cmol_c kg^{-1}$), yield (28-38%), surface area (159-246 m^2/g), pore diameter (2-19 nm) and pore volume (0.01-0.6 cm^3/g) respectively[13].

5.3.2 Intermediate Pyrolysis

Intermediate pyrolysis has acquired significant attention in recent years, due to satisfactory raw biooil production[14] (up to 50 % yield from woody biomass), with less residence time from 10 to 30 s, at a temperature nearly 500 °C. Although, this kind of pyrolysis allows making use of larger particle size of feedstock (including chips and pellets) while in case of fast pyrolysis required finely ground raw material. Therefore, this process considered as most robust and reliable because of its suitability in small and medium scale industries.

Funke et al.[15] compared intermediate and fast pyrolysis on the basis of different biomass feedstock (hybrid poplar, beech wood and wheat straw) by keeping same pyrolysis temperature (500 °C). Authors recorded minimum organic liquid yield (17-21 %) from intermediate pyrolysis as compared to fast pyrolysis 26-51 %. The lower biooil yield for intermediate pyrolysis leads to more formation secondary products including char, pyrolysis syngases, and reaction water. The higher biooil yield from fast pyrolysis was decreased in ash content of biomass due to catalytically active minerals. Mohammed et al.[16] conducted intermediate pyrolysis in vertical fixed bed tubular reactor employed by Napier grass to analyze the effect of process temperature, heating rate and nitrogen flow rate on resulting end products. Authors recorded optimum biooil yield of 50.57 wt% at a temperature of 600 °C, 50 °C/min heating rate and 5 l/min nitrogen flow rate, further they characterized biooil, biochar and syngases using standard analytical techniques. The organic phase oil consists of benzene derivatives such as phenols, ethyl-phenols, methyl-phenol, methoxy-benzene, methoxy-phenols etc. which constituted about 60% of the total organic phase, hence it showed that the strong agreement of biooil composition. However, produced biochar from intermediate pyrolysis process showed porous carbonaceous organic material, rich in mineral composition, which may be used as a solid fuel due to high energy content, as an adsorbent in wastewater treatment, macronutrient for agriculture. The syngas composed of hydrogen, carbon dioxide, carbon monoxide, and methane. The main components in gas were observed at temperature 450 °C as carbon dioxide (16.42 vol%), hydrogen (15.44 vol%) and carbon monoxide (9.79%) respectively.

5.3.3 Fast Pyrolysis

Fast pyrolysis of biomass for renewable biofuel production is considered as one of the cheapest conversion routes. In fast pyrolysis process thermal degradation of biomass takes place in absence of air/oxygen at a moderate temperature (450-600 °C) with a very short residence time (< 2s). The major resulting end product in fast pyrolysis is an initial dark brown colored viscous liquid known as biooil or pyrolytic oil followed by biochar and syngases. The fast pyrolysis process has maximum conversion efficiency and a recorded highest yield of biooil about 75 % on a dry mass basis[17]. As compared to other biofuel production technologies, biooil produced through a fast pyrolysis process is considered a cost-effective thermochemical conversion route. The profitability of the fast pyrolysis process depends on parameters such as product yield, quality of product, feedstock cost, production scale etc. The fast pyrolysis reactors such as a bubbling fluidized bed, rotating cone, screw/auger, ablative, and vacuum pyrolysis are considered suitable technologies for biooil production, although the configuration of fluidized bed and rotating cone reactors are viewed as the most cost effective and commercialized technologies of the group.

In a fast pyrolysis process, the syngas yield has a badly affect on biochar and the biooil yield because as syngas yield increases, the yield of biochar and biooil drops suddenly. The biooil produced from woody biomass shows a higher yield than agricultural by-products, forest residue, and energy crops etc. because clean woody biomass contains a high percentage of cellulose and hemicellulose which is more favorable for biooil production. The selection of feedstock particle size in fast pyrolysis process can adversely affect on biooil, char and syngas yield, because as used particle size is more, heat transfer rate slightly decreases so it causes raising the biochar yield, while drop in both biooil and syngas yield. Therefore, for getting the maximum yield of biooil small particle size feedstock should be preferred. Xue et al.[18] carried out fast pyrolysis experiment by using mixture of biomass and waste plastic in a fluidized bed reactor at a temperature range between 525 to 675 °C. during the experiment the maximum yield of biooil was recorded about 57.6 % at 625 °C temperature and it was observed that the due to co-pyrolysis resulted biooil had a good quality and having higher heating value was about 36.6 MJ/kg.

5.3.4 Flash Pyrolysis

In flash pyrolysis process, thermal degradation of biomass take place at higher heating rate (from 10^3 to 10^4 °C/s) by keeping a very short residence time (< 0.5s), resulted a higher biooil yield (75-80 wt%). If maximum biooil production is goal then flash/fast pyrolysis is more recognized process as compared to other. Generally, flash pyrolysis reaction take place with a fraction of seconds but it requires a higher heating rate so due to instant heating, biomass particle size should be small. To achieve the higher biooil yield (75%wt.) at a higher temperature range from 800 to 1000 °C, then feedstock particle size should be less than 200 micron meter[19]. Owing to this, Raja et al.[20] used jathropha oil cake as a feedstock having a particle size ranged between 0.6-1.18 mm in a fluidized bed reactor for production of biooil through flash pyrolysis process. It was observed that, maximum biooil yield was obtained when feedstock particle size was 1.0 mm.

Various reactor configurations such as fluidized bed, rotating cone, circulating fluidized bed, ablative, vacuum moving bed reactors, and others are more suitable for biooil production through flash pyrolysis. Fluidized bed reactor gives a higher yield of biooil (nearly 70 %) by using soyabean flake as a feedstock at 550 °C and requires very short residence time (0.3 to 0.6 s)[21]. Apart from this, rotating cone reactor is also considered a good configuration for flash pyrolysis process due to rapid heating rate and short residence time[22]. Makibar et al.[23] conducted a flash pyrolysis experiment in a conical spouted bed reactor by using poplar (*Populus nigra*) as a feedstock and recorded biooil yield in two fractions at a temperature range between 425-525 °C. Authors observed that in first fraction biooil yield was about 85 % wt and contained a higher proportion of water, while another fraction gave a biooil yield (25 % wt) with lower water content. In addition, authors also noticed that produced biooil with a low water content had a higher heating value (HHV) in the range of 16-18 MJ/kg respectively.

5.4 Properties of Pyrolysis Products

5.4.1 Properties of Bio-oil

Bio-oil, an organic mixture with many components, which obtained by the fragmentation and depolymerization of biomass that contains hemicelluloses, cellulose, and lignin. The properties of bio-oil depend mainly on two factors: (1) type of selected biomass or feedstock and (2) operating conditions. Bio-oil is a dark red-brown to almost black liquid and its colour composition changes with its chemical properties and micro carbon content. The chemical species in the biooil play significant role for determining its quality, suitability and stability towards the further upgrading. The pyrolytic oil termed biooil is composed of numerous organic compounds (it may be several hundred), which exhibits good chemical functionalities. Therefore, biooil can be used either as a fuel or as a valuable chemical. Oyebanji et al.[24] analyzed chemical species in biooil produced from fast pyrolysis of west African cordia and African birch sawdust as energy biomass. Authors analyzed different phenolic compounds, aromatic hydrocarbons, oleic acid, and nitrogen-containing compounds. The level of oleic acid and phenolic compounds was higher in biooil derived from cordia biomass than birch sawdust. The aromatic hydrocarbons present in biooil are mainly indene, benzene, naphthalene, tolyene, and methylnaphthalene. The availability of phenolic compounds in biooil subjected to replacement for fossil phenol for the preparation of different chemicals. The biooil produced from mango waste (tegument and almond) at 650 °C temperatures in a fixed bed reactor and observed the more availability of phenols (32.6%) and ketones (22.9 %) in biooil derived from tegument and in contrast availability of ketones (20.6%), hydrocarbons (7.2 %) and acid (16.8%) in almond derived biooil[25].

5.4.2 Properties of Biochar

The partial or complete decomposition of biomass at elevated temperature in absence of air resulted in primary solid end product known as biochar or charcoal. The physicochemical properties of biochar significantly vary according to pyrolysis condition and type of feedstock. Generally, slow pyrolysis process widely adopted for getting a higher yield of biochar (35 %) followed by gas (35%) and biooil (30 wt%) at a pyrolysis temperature (300-800 °C) for longer duration (maybe hours). Wang et al.[26], summarized the elemental and physicochemical composition of biochar from pyrolysis of different agrowaste at different heating rate and pyrolyzing temperatures (300-600 °C) as, carbon-containing (60-80%), N (0.3-3.10), P (0.03-0.3), K (0.02-1.21), O/C atomic ratio (0.1-0.4), H/C atomic ratio (0.3-0.8), pH (7-10), BET surface area (0.1-16 m^2/g) and pore volume (0.002-0.02 ccg^{-1}) respectively. All the produced biochar were rich in organic carbon, which was due to as pyrolysis temperature increased, and thus the organic carbon content in biochar also improved, while oxygen and hydrogen content were significantly reduced. Also decrease in H/C and O/C ratio of all biochar at higher temperature indicated loss of oxygen-containing functional groups and formed graphic as well as aromatic structure. pH of biochar may be influenced by pyrolysis temperature, as pyrolysis temperature increased more alkaline cations (Ca, Mg and k) accumulate on biochar surface which causes increase in pH. The surface area, pore-volume, i.e. the microscopic surface structure of biochar after pyrolysis were showed a good potential ability for adsorption and filtration of organic and inorganic pollutants[27]. The higher heating value of biochar (15-30 MJ/kg), varies according to pyrolysis condition and feedstock[28]. The maximum heating values of biochar become an attractive alternative to coal in fuel application.

5.4.3 Properties of Syngas

Syngas or pyrolytic gas are the gases released during biomass pyrolysis it may be combustible or non combustible as carbon dioxide (CO_2), hydrogen (H_2), carbon monoxide (CO), methane (CH_4), ethane (C_2H_6), among which H_2 and CH_4 considered as energy fuels and also some other gases including ammonia (NH_3), nitrogen oxide (NOx), propane (C_3H_8), sulphur oxide (SOx) etc. The primary gas constituents CO_2 and CO originate from biomass pyrolysis and reforming the carboxyl (CaO) and carbonyl (C=O) groups.

Typically the calorific value of pyrolytic gases ranged from 10-20 MJ/Nm^3 depending on the pyrolytic conditions, among the pyrolytic conditions pyrolysis temperature significantly affect on yield and composition of syngas. Biomass composed of some macro-molecular components, which couldn't completely decompose at low pyrolytic temperature; otherwise, there is breakage of weak chemical bond. Therefore as temperature increased condensation polymerization take a good position with gradual increases in gas production. More amount of CO_2 is formed during pyrolysis process is due to the cracking and reforming reaction of carboxyl and carbonyl group for organic compound and carbonate decomposition for inorganic compound. Therefore, more CO_2 formation indicated that high oxygen content present in biomass. Ábrego et al.[29] analyzed syngas composition for pyrolysis of cashew nutshell at 500 °C temperature as CO_2 (11.65%), CO (2.08%), H_2 (0.09%), CH_4 (2.57%), C_2H_4 (0.13%), C_2H_6 (0.55%) and HHV of gas was about 11.9 MJkg^{-1} respectively. The methane is mainly combustible gas released due to cracking of methyl (-CH_3), methoxy (-O CH_3), and methylene (-CH_2-) groups at higher pyrolysis temperature[30]. Bensidhom et al. [31] carried out pyrolysis of date palm waste in a fixed bed reactor at 500 °C, with a 15 °C/min as a heating rate and recorded syngas yield in between 39-46 % respectively. Meanwhile, in the experiment authors observed that monoxide carbon (31-57%) and dioxide carbon (0.33-0.55%) are the dominant compounds present in syngas. The hemicelluloses and cellulose material showed higher CO and CO_2 yield, while lignin-derived syngas showed more CH_4 releasing. Actually hemicelluloses and cellulose biomass contain more availability of oxygen as compared to lignin, therefore more formation of CO and CO_2. Lignin biomass owns more hydrogen and methane yield, it might be due to availability of higher O-CH_3 functional groups and aromatic rings in lignin material, hydrogen is mainly formed due to cracking and distortion of C=C(ar) and C=H (ar) groups[32].

5.5 Upgrading of Bio-oil

Due to its application in industrial, transportation, and commercial sector, the up-gradation of bio-oil has been become a research hotspot. Compared to fossil fuels such as diesel, petrol, and other, bio-oil has high water content, acidity, corrosiveness, density and viscosity, which limits its direct use in many applications. As such researchers are faced with the challenges of improving the combustion performance of the bio-oil by reducing its corrosion rate and increasing its H/C ratio. To enhance the quality of bio-oil, a number of mechanisms can be considered, including physical, chemical, and catalytic upgrading methods, as summarized and shown in Fig. 5.2

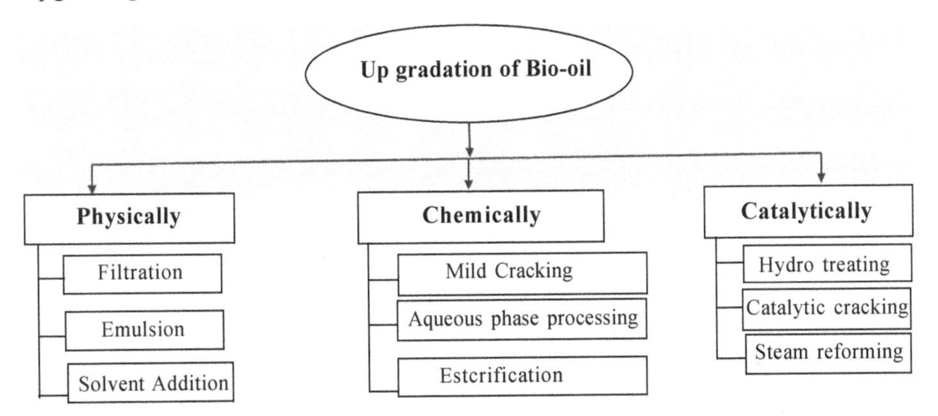

Fig. 5.2: Classification of Upgrading of Bio-Oil

5.5.1 Physically Up Gradation of Bio-oil

Physical upgrading of bio-oil is done by various methods including Filtration, Emulsion and Solvent addition as explained subsequently.

5.5.1.1. Filtration

Hot vapour filtration (HVF) is a type of bio-oil physical upgrade process in which pyrolysis vapour are used for filtration prior to condensation process. HVF play important role in reducing the both ash content (< 0.01 %) and the alkali content (< 10 ppm) in resulting bio-oil, as reported by Bridgwater[17]. Hot vapour filtration is non catalytic pyrolysis process, in which hot gas filters are made from material which is compatible to acidic properties of oil. Hot vapour filtration, gives good quality biooil, however too much quantity (10-30%) of liquid yield was lost in filtration process, because of catalytic reaction and plugging problem in filter.

Similarly, the average molecular weight and alkali metal content was reduced by increasing the water content and the pH of the bio-oil after HVF. On the other hand, the HVF method led to a reduction in bio-oil yield (6-7 %), while on the other hand, the filtered oil showed better qualities in terms of stability, ash content, solid content, and viscosity as reported by Pattiya and Suttibak[33]. However, in HVF as it has some limitations, if the particle size of the bio-oil is less than 5-micron meters, then it will be very difficult for filtration due to its physic-chemical nature and also requires a self-leaning. Advantages and limitations of hot water vapour filtration process are listed in Table 5.1.

Table 5.1: Advantages and Limitations of Hot Water Vapour Filtration

Type of filtration	Advantages	Limitations
Hot vapour filtration (HVF)	• Reduces the ash (< 0.01 %) and alkali metal (< 10 ppm) content in bio-oil. • Gives higher quality of product with a lower char • Reduces pH of bio-oil. • Filtered oil shows better qualities in terms of stability, viscosity, ash content, solid content etc.	• Lower molecular weight of bio-oil affects engine performance. • Minimizes the yield of bio-oil. • Less particle size bio-oil affect the filtration. • Need of self-cleaning filter.

5.5.1.2. Solvent Addition

To increase the heating value and reduce the viscosity of bio-oil polar solvents like methanol, ethanol and furfural (having a higher heating value) have been used from many years. Correspondingly, polar solvents have an ability to homogenize the bio-oil. Many studies have revealed that the direct use of solvents after pyrolysis creates a significant effect on the stability of bio-oil, they increases stability and the heating value of the oil. Mei et al.[34] reported that methanol was added into pinewood biooil at various proportions (3, 6, 9, 12, 15 wt %) respectively and studied different physiochemical properties including pH value, water content, and viscosity for a 35 days of storage. Water content and viscosity were reduced, on the other hand methanol content was raised 6 wt%, and pH value was also significantly increased. Finally, authors concluded that addition of methanol (solvent addition) into biooil aids to improve the storability and creates a favourable condition for further application.

5.5.1.3. Emulsion

Biooil having some unpleasant physiochemical properties as discussed earlier, this restricts its direct application as a transportation biofuel. Therefore, currently emulsion of biooil and disel considered as a promising option for upgrading of oil and its further efficient application as a transport fuel. Farooq et al.[35], carried out emulsion of ether extracted biooil (EEO) with diesel fuel by addition of emulsifier for 40 days of duration at room temperature. The term hydrophilic-lipophilic balance value (HLB) is very important, which varies according to chemical structure and properties of emulsifier. Therefore, it could be considered as a key parameter for selection of surfactant. Here, authors measured a HHV of emulsified fuel which were 44.32 and 43.68 MJ/kg (nearly equal to diesel HHV 45 MJ/kg), whereas emulsifier mass ratio (EEO: Emulsifier) were 5:5 and 7.4:6.6 respectively.

The University of Florence has performed an experiment on emulsion of 5-95 % of bio-oil in the diesel either for diesel engines for the generation of power or as a transport fuel to operate the diesel engine on dual fuel mode.

5.5.2 Chemically Upgrading of Bio-oil

Chemical upgrading of bio-oil is done by different methods including aqueous phase processing/reforming, mild cracking and esterification as explained subsequently.

5.5.2.1 Aqueous Phase Processing/Reforming

Aqueous phase processing is one of the optimistic option for the conversion of biomass into bio-fuels especially H_2 and alkane. During the study they found that Pt is a good monometallic catalyst in terms of activeness and choice for aqueous phase processing.Vispute and Huber[36] produced hydrogen, alkanes (C_1-$C_{6)}$ and polyols from aqueous phase processing of wood-derived bio-oil. The methodologies adapted during aqueous phase processing are illustrated in Fig. 5.3.

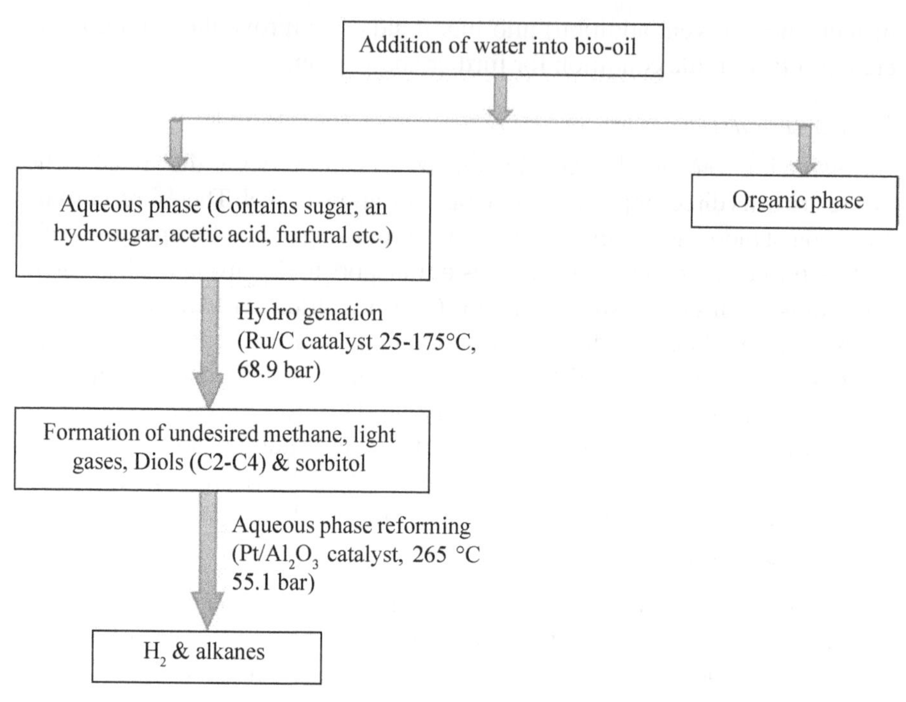

Fig. 5.3: Flow Chart of Aqueous Phase Reforming

5.5.2.2 Mild Cracking

An alternative option to acid catalyzed cracking or hydro treating process is mild cracking, while using acid cracking catalyst like H-ZSM-5 results higher yield of gases hydrocarbon as well as formation of coke, which affects the performance of the catalyst; on the other hand, the main aim of mild cracking is to reduce coke and gas formation as studied by Fisk et al.[37] because in the mild cracking reaction, oxygen was partly removed from the bio-oil, thus avoiding the problem.

Nokkosmaki et al.[38] used micro and bench scale reactor for upgrading of pyrolysis vapours using Zinc oxide catalyst, and they proved that ZnO was promising mild catalyst plays a significant role without affecting the liquid yield.

5.5.2.3 Esterification

Bio-oil has some drawbacks, which include poor stability, high viscosity, low heating value, high acidity, and corrosiveness. These properties restrict bio-oil's high-grade applications. Consequently, for further use in a wide range of applications, there is a need to upgrade the bio-oil. Bio-oil consists of two major compounds: inorganic matter and organic acid. Organic acid is comprised of many acid types, e.g., acetic acid, formic acid, propionic acid. The organic acid can be converted into esters during the esterification process, which can enhance the quality of the bio-oil in terms of HHV and stability.

Wang et al.[39] upgraded bio-oil by using the 732- and NKC-9-type ion exchange resins as the selected catalysts. The results have shown that acid numbers are reduced by 88.54 and 88.95%, HHV increased by 32.26% and 31.64%, the moisture content significantly decreased by 27.74% and 30.87%, and densities and viscosities decreased by 21.77% and 97%, respectively.

A solid acid catalyst, ionic liquid catalyst, solid base catalyst, HZSM-5, and aluminum silicate catalyst can also be used in the esterification process.

5.5.3 Catalytically Up Gradation of Bio-oil

Catalytic upgrading of bio-oil is done by different methods including hydro treating, catalytic cracking and steam reforming as explained subsequently.

5.5.3.1 Hydro-treating

It is well-known that a fuel that contains higher percentage of hydrogen has better ignition quality. Presently, hydrotreating is the most effective technique for upgrading bio-oil by removing the oxygen. Hydrogenation involves simply adding hydrogen in a refinery. Hydrotreatment is one of the effective techniques for producing stable compounds from aldehydes and saturated compounds by using a catalyst, but the conversion process requires more temperature and hydrogen pressure to react with acid, which is economically viable and energy inefficient studied. There are two catalysts commonly used in the hydrotreating process sulphide $CoMo/Al_2O_3$, and $NiMo/Al_2O_3$.

Zhang et al.[40] conducted an experiment by using the hydrotreating technique in presence of Ni/HSZM-5 catalyst to upgrade bio-oil obtained from pyrolysis of saw-dust and reported an increase in pH from 2.27 to 4.07 correspondingly; hydrogen content increased from 6.28 to 7.01 wt.%; and gross calorific value from 13.79 to 14.32 MJ/kg. There are some limitations in the hydrotreating process, such as it requires mild conditions for operation and low yields of upgraded bio-oil, and this technique produces a large amount of char and tar that causes clogging problems in reactor. Benefits and drawbacks of bio-oil upgrading through different methods are listed in Table 5.2.

Table 5.2: Benefits and Drawbacks of Bio-oil Upgrading

Method of upgrading	Operating condition and catalysts	Benefits	Drawbacks
Hydrotreating/ Hydrorifining/ catalytic hydrogenation/ Hydrodeoxyg-enation (HDO)s	• Mild condition • Operating temperature range from 250 – 350 °C and pressure 100-200 bar. • $NiMo/Al_2O_3$, $CoMo/Al_2O_3$ [204]. • Palladium or ruthenium catalyst. • Ni/HSZM-5 catalyst, PdSZr catalyst.	• Most effective technique by removing the oxygen. • Increase in pH, hydrogen content and HHV. • Reduces viscosity as well as density. • Used as transportation fuel.	• More expensive because it requires H_2. • Necessity of high pressure (70-200 bar). • Low yield of upgraded bio-oil and also produce char, tar which creates problem in clogging reactor.

5.5.3.2 *Catalytic-cracking*

Catalytic cracking is the most promising technique for the upgrading of bio-crude in to liquid fuel. Catalytic cracking helps to remove the oxygen as water and carbon oxide from crude oil by using a zeolite catalyst. Basically, there are two main types of catalytic cracking as classified in Fig 5.4.

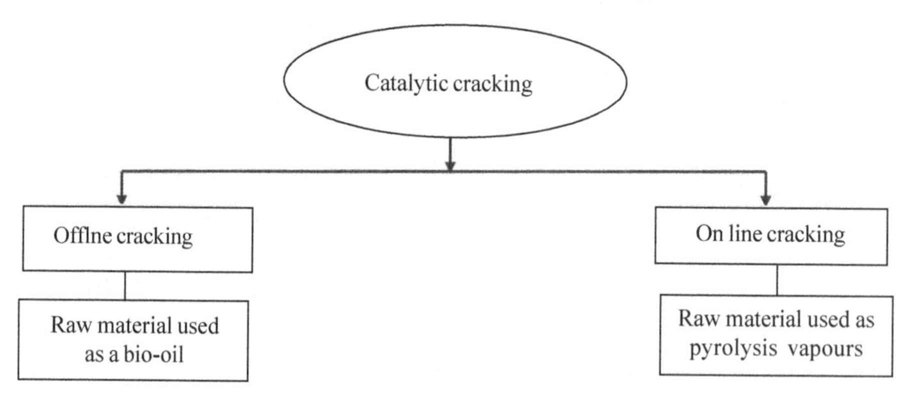

Fig. 5.4: Classification of Catalytic Cracking

Hew et al.[41] carried out a study on conversion of bio-oil into gasoline fuel by adopting a catalytic cracking process in presence of a Zeolite ZSM-5 catalyst at a temperature of about 400 °C for 15 minutes and reported that the yield of gasoline was about 91.67 wt%. Padamaja et al.[42] carried out a catalytic cracking process for the conversion of bio-crude into upgraded liquid fuel in a continuous mode micro reactor at a temperature varied from 460 - 520 °C. They observed that maximum conversion was made into gas and liquid products. The gas (68-

92 %) yield was recorded and liquid fuel yield ranged from 59 - 78%. Cheng et al.[43] performed catalytic cracking of crude oil by using zeolite catalyst and they noted the higher heating value of upgraded bio-oil was about 41.4 MJ/kg, as compared with crude bio-oil having 30.54 MJ/kg.

5.5.3.3 Steam Reforming

Today, a catalytic-steam-reforming technique is used to produce hydrogen from bio-oil. A major benefit of producing hydrogen with the steam reforming of bio-oil is that it becomes less expensive and much easier than other fuels. Thus, bio-oil steam reforming has the ability to produce renewable, clean, and clear hydrogen along with an upgraded bio-oil which has a remarkable impact on the generation of power using fuel cell and for transportation. Recently Al-Rahbi and Williams[44] used waste ash as (ash from coal, tyre and reused derived fuel etc.) a catalyst for pyrolysis of biomass, followed by catalytic steam reforming to enhance the hydrogen gas production. Here, authors observed that , due to presence of ash catalyst, yield of H_2 gas significantly increased and better yield recorded for reused derived fuel ash notably 7.90 mmolg^{-1} biomass. The reason for increase in hydrogen gas yield was the availability of metal content (Al, Cu, Mg and Fe, Na, Zn etc.) in derived ash sample; therefore theses metals acted as a catalyst and help to enrich the hydrogen gas production.

Generally, a fixed or fluidized bed reactor is preferred for the steam reforming process and it requires a maximum temperature in the range of 800 to 900 °C. Similarly, Garcia et al.[45] used a fixed bed micro reactor for the production of hydrogen in steam reforming process of condensable vapours, with a reactor temperature kept at about 825-875 °C, high space velocity (126,000 h^{-1}) and minimum residence time (26 ms). Results showed that catalytic-steam reforming efficiency depends on water-gas shift activity in the catalyst. The selection of a good catalyst plays an important role in the steam reforming process. Sometime, the formation of coke deactivates the catalyst, which begins at an operating temperature[46] of 575-900°C. The formation of carbon can be eliminated by increasing the steam to carbon ratio.

5.6. Applications of Pyrolysis Products

5.6.1 Applications of Bio-oil

The bio-oil produced from different pyrolyzing techniques have thermal, commercial, and industrial applications (Fig. 5.5). It can be used for combustion to generate heat and power, binding agent to make pellets and briquettes etc. However, properties of bio-oil such as poor volatility, high viscosity, coking and corrosiveness create significant problems in direct use or applications in furnaces, boilers, engines, gas turbines, chemicals, etc. Some minor modifications and additions are required in burners, engines, boilers, etc. to optimize the maximum combustion efficiency.

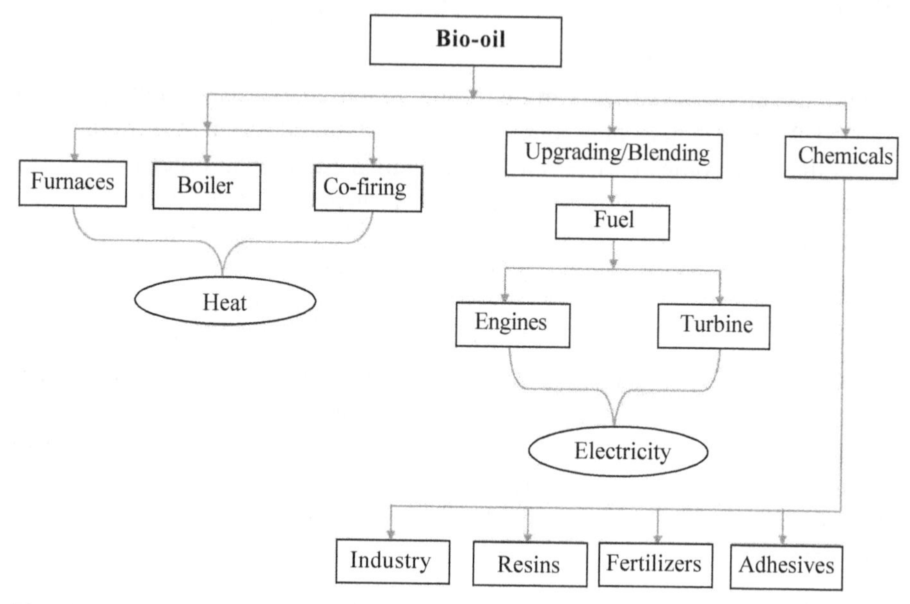

Fig. 5.5: Various Applications of Bio-oil

5.6.1.1 Furnace/Boiler/Co-firing

The biomass derived fuel as bio-oil is used to generate heat and power instead of using fossil fuels like diesel in stationary applications such as boiler, furnace etc. The furnaces, boiler can operate with a wide range of fuels but less competent than engine and turbine.

Existing industrial boiler can be run on biooil by replacing the fossile fuel is possible by few modifications in existing combustion chamber in order to enhance the combustion stability. As biooil can be co-fired with a conventional fossile fuels for exploring the advantage of maximize the overall efficiency by reducing the emission of CO_2, SO_x and NO_x

5.6.1.2 Diesel Engine

The boiler and furnace are commonly used to produce the heat, but diesel engine generates power at a maximum efficiency (45%), and it can also adapt in a cycle that generates both heat and power. Bio-oil can be used at a slow and medium speed of a conventional diesel engine; however, sometimes it creates a problem when used in the engine application directly. These problems included the deposition of carbon on piston as well as on components of engine combustion chamber, gum and wax formation, injector coking, poor atomization, engine wear, and starting difficulty in cold weather, among others. Recently, Yuan X et al.[47] were combusted biooil emulsion with diesel in Direct Injection (DI) diesel engine for investigation of biooil properties in engine performance and to check the emission index. They observed that as compared to diesel, synthetic biooil causes increase in break specific fuel consumption (BSFC) and energy consumption, although break power remained same for emulsion. Authors also concluded with respect to emission level, carbon monoxide and smoke emission were dropped, while carbon dioxide emission was slightly increased. In addition, authors also noticed that emulsion obtained from hydroxyle compounds and aldehydes significantly reduced NO_x emission, while emulsion made from sugar gas raised NO_x emission.

In the Compression Ignition (CI) engine 50-80% of rubber seed oil can substitute for diesel fuel easily without any changes, modifications, engine structure and operational difficulties, as studied[48]. During the experiment, they tested performance and emission level of the CI engine using a blend of rubber seed oil and diesel fuel. They found that the blend-fueled CI engine had more carbon deposition inside the combustion chamber as compared to a diesel-fueled engine. Van de Beld et al.[49] studied the effect of different parameters on performance of a one-cylinder 20 kW diesel engine for power generation, installed using a stainless-steel fuel injection system and injector to enable the fueling of bio-oil and upgraded bio-oil. They found very promising results during the experiment when they filled the pyrolysis oil in the diesel engine for duration of 40 h and found that there was a notable effect on fuel consumption as well as flue gas emissions. They also estimated the cetane number of pyrolysis oil in the range of 20-25 and reported that pyrolysis oil is easily accessible to ignite in the engine as compared to butanol and bio-ethanol. As noticed in many studies have reported various technical challenges associated with the application of bio-oil in heat and power generation, such as deposition of coke, ash, or carbon in the combustion chamber, poor ignition quality, injector clogging and polymerized material and corrosion. In addition, Costro et al.[50] carried out fractional distillation of bioooil produced from pyrolysis of lignin cellulose residue at 450°C temperature. Further, authors also reported the distillation biooil obtained yield of fossile fuel fractions as gasoline (4.70%), kerosene (28.21%), and light diesel oil (22.35%) respectively, which shown hydrocarbon rich mixture for further energy fuel application.

5.6.1.3 Gas Turbine

Gas turbines are extensively used in power plant to generate electric power, which operates on petroleum fuels. It can be modified to make them compatible with bio-oil. Owing to this, Buffi M et al.[51] checked the performance of micro gas turbine by using fast pyrolysis biooil blends with ethanol, included a modified combustor and new fuel injection line. The demonstration of micro gas turbine for a test as 20:80 and 50:50 (volume fraction) of biooil blend with ethanol performed a stable and efficient engine operation. Authors are also observed that engine performed an overall electrical efficiency more than diesel fuel, it occurred due to redesigned combustor and fuel line. In addition, authors also highlighted the emission status, as biooil volume fraction was increased; carbon monoxide emission happened very rapidly, probably it occurred due to high viscous biooil along with a large droplet size. Finally they also checked performance on 100 % biooil, found as unstable fuel combustion, deposition of carbon particles, showing that it required further modifications to reach at desired goal.

The testing of the first gas turbine J69-T-29 using biomass-derived pyrolysis oil was carried out at Teledyne CAE (USA)[52]. During the testing, the combustion efficiency of pyrolysis oil as a fuel was reported about 95 %, but it was expected that it would exceed up to 99% in the engine at the optimum condition.

5.6.1.4 Chemicals

The bio-oil derived from the pyrolysis process has the potential to produce high-grade chemicals that are used to make fertilizers, acetic acid, food flavourings, phenol, and sugars, etc for industrial applications. Bio-oil is a complex mixture of different compounds including phenols, acids, ketones, aldehydes, hydrocarbon, sugars, ethers, ester etc.Therefore, biooil has a potential to produce different chemicals for further industrial application, e.g. levoglucosan for the preparation of pharmaceuticals, phenolic compounds used in phenolic resins, surfactant, biodegradable polymer, and hydroxyacetaldehyde, considered as meat browning agents.

5.6.2 Applications of Biochar

Biochar is a stable organic carbon-rich product obtained from thermal decomposition of biomass in the absence of air or limited oxygen environment. There are different applications of biochar which especially covers four synergistic intentions; waste management, soil improvement, energy production, and climate change mitigation, shown in Fig 5.6. Biochar produced at a pyrolysis temperature range between (350-600 °C) has shown widespread applications in the agriculture sector, especially it acts as a soil amendment. It also helps in improving soil quality and enhances nutrient availability. However, a soil amendment quality of biochar significantly varies with feedstock source and pyrolysis temperature. On the other side, biochar produced at high pyrolysis temperature (400-800 °C) is mainly composed of aromatic carbon and it can be considered as carbon sequester for soil application. Although, there are few limitations to use as a soil amendment due to high pyrolysis temperature produced biochar possesses low ion-exchange functionality. Therefore, it might be used as a bio sorbent for adsorption of heavy metals due to abundance availability of organic functional groups on the biochar surface and inorganic minerals[53]. Biochar attributed a great potential in sustainable energy applications including supercapacitors, fuel cells, and new battery technologies[54] etc. As we discussed earlier, biochar produced at high pyrolysis temperature showed high surface area, porosity, electrical conductivity, etc. which signify the most appropriate characteristics for future use as a supercapacitor electrode. Biochar obtained from woody biomass can be optimized potentially in direct carbon fuel cell for electric energy generation. However, sometimes it gives lower performance, due to low carbon content and high ash content present in biochar. Biochar after activation (activated biochar) may be used in most popular lithium-ion batteries, in which graphine is considered as well known anode material.

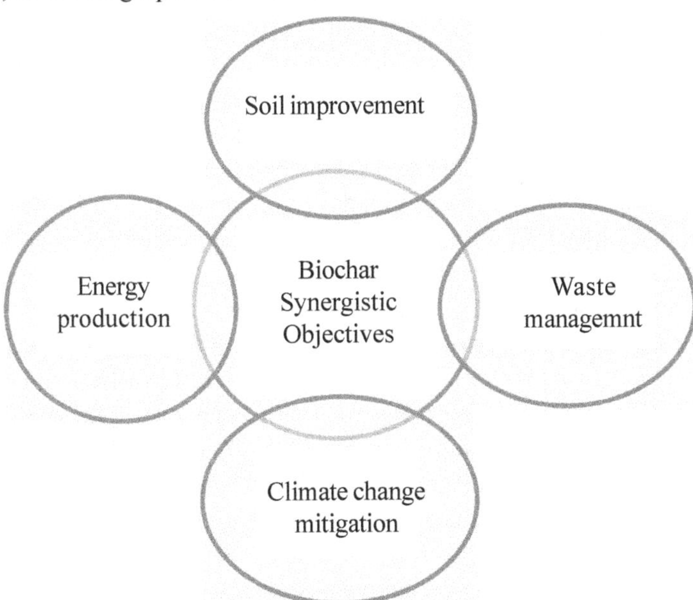

Fig. 5.6: Synergistic Objectives of Bio-char

5.6.3 Applications of Syngas

In the case of pyrolysis process, thermal degradation of biomass takes place results in formation of both condensable gases (vapours) and non condensable gases (primary gases). The vapours composed of heavier molecules as discussed earlier condenses and adding biooil yield. While, non condensable gases, which are made of low molecular weight gases such as carbon monoxide (CO), carbon dioxide (CO_2), hydrogen (H_2), methane, ethane etc. which is also called as ' primary gases', 'syngases', 'biogas', 'pyrolytic gases'. Theses syngas having some heating value (11-20 MJ/Nm^3) therefore these gases had a potential application, as directly utilized for heat or electricity generation (either gas combustion in SI and CI engines). Owing to this, Gupta et al.[55] developed syngas fired combustor which is giving stable performance, high turndown ratio and quick start-up for solar thermal application. In some cases, syngas can be co-fired with coal, for getting methane, hydrogen energy-rich gas or in preparation of liquid biofuel using synthesis technique[56]. In some application, syngas might be used as carrier gas in pyrolysis reactor or also used for preheating the inert sweeping gas. In addition, syngas is considered an important source for the preparation of valuable chemicals; fertilizers through ammonia, methanol comes from syngases used for chemical industry etc.

5.7. Potential of Greenhouse Gas Emission

Biomass is considered as prime renewable energy sources, which can be utilized to produce electric power, process heat for industrial applications, and liquid fuel for transportation purposes. Biomass energy helps in reducing greenhouse gases compared to fossil fuels. The International Energy Agency (IEA)[6] projected about almost 20% of the cumulative carbon savings through biomass energy by 2060. Shemfe e al.[57] examined the greenhouse gas emission potential with upgraded bio-oil produced with fast pyrolysis reactor and upgraded via hydro-processing and zeolite cracking. Further, they compared the bio-hydro produced from these two upgrading routes and fossil fuel and found that there is saving of greenhouse gas emission between 68% and 87%. Life cycle assessment (LCA) used for for fuel produced through fast pyrolysis and upgraded by hydrotreating and hydrocracking to produce gasoline and diesel. Hsu[58] reported that greenhouse gas emission of CO_2 equivalent of 117 g km^{-1} and net energy value of 1.09 MJ km^{-1} for pyrolysis derived gasoline and for diesel derived through the same process have a CO_2 equivalent of 98 gkm^{-1} and net energy value of 0.92 MJ km^{-1}. Further, it was also reported that pyrolysis derived gasoline and diesel have lower greenhouse gas emissions and higher net energy value than conventional gasoline in 2005.

References

[1] Goyal H-B, Seal D, Saxena R-C. Bio-fuels from thermochemical conversion of renewable resources: A review, Renew Sustain Energy Rev. 2008;12: 504–517.
[2] McKendry P. Energy production from biomass (part 1): overview of biomass, Elsevier. 2002; 83: 37–46.
[3] Gercel H-F. The production and evaluation of bio-oils from the pyrolysis of sun ower-oil cake, Biomass and Bioenergy. 2002; 23: 307–314.

[4] Aysu T, Durak H, Güner S, Bengü A-S, Esim N. Bio-oil production via catalytic pyrolysis of Anchusa azurea: Effects of operating conditions on product yields and chromatographic characterization. Bioresource Technol. 2016; 205: 7-14.

[5] Czernik S, Bridgwater A-V. Overview of applications of biomass fast pyrolysis oil. Energy & fuels 2004; 18(2): 590-598.

[6] IEA. "Technology Roadmap; Delivering Sustainable Bioenergy," International Energy Agency, Paris, France 2017.

[7] Abnisa F, Arami-Niya A, Daud W-W, Sahu J-N, Noor I-M. Utilization of oil palm tree residues to produce bio-oil and bio-char via pyrolysis, Energy conversion and management 2013; 76: 1073-1082.

[8] Ministry of new renewable energy, govt of India, https://mnre.gov.in/biomass-powercogen assessed on 26 August, 2019

[9] Chaitanya B, Bahadur V, Thakur A-D, Raj R. Biomass-gasification-based atmospheric water harvesting in India. Energy: 2018165: 610-621.

[10] Marshall A-J (2013) Commercial application of pyrolysis technology in agriculture.)#www.ofa.on.ca/uploads/userfiles/files/pyrolysis%20report%20final.pdf*#

[11] Hornung A. Transformation of biomass: theory to practice. John Wiley & Sons 2014.

[12] Yu S, Park J, Kim M, Ryu C, Park J. Characterization of biochar and byproducts from slow pyrolysis of hinoki cypress. Bioresource Technology Reports 2019.

[13] Mandal A, Singh N, Purakayastha T-J. Characterization of pesticide sorption behaviour of slow pyrolysis biochars as low cost adsorbent for atrazine and imidacloprid removal. Science of the Total Environment, 2017; 577: 376-385.

[14] Torri I-D-V, Paasikallio V, Faccini C-S, Huff R, Caramão E-B, Sacon V, Zini C-A. Bio-oil production of softwood and hardwood forest industry residues through fast and intermediate pyrolysis and its chromatographic characterization. Bioresource technology, 2016; 200: 680-690.

[15] Funke A, Morgano M-T, Dahmen N, Leibold H. Experimental comparison of two bench scale units for fast and intermediate pyrolysis. Journal of Analytical and Applied Pyrolysis, 2017; 124: 504-514.

[16] Mohammed I-Y, Abakr Y-A, Yusup S, Kazi F-K. Valorization of Napier grass via intermediate pyrolysis: Optimization using response surface methodology and pyrolysis products characterization. Journal of cleaner production, 2017; 142: 1848-1866.

[17] Bridgwater A-V. Review of fast pyrolysis of biomass and product upgrading, Biomass and Bioenergy 2012; 38: 68-94

[18] Xue Y, Zhou S, Brown R-C, Kelkar A, Bai X. Fast pyrolysis of biomass and waste plastic in a fluidized bed reactor. Fuel, 2015; 156: 40-46.

[19] Balat M, Balat M, Kirtay E, Balat H. Main routes for the thermo-conversion of biomass into fuels and chemicals. Part 1: pyrolysis systems. Energy Convers Manag 2009; 50: 3147–57.

[20] Raja S-A, Kennedy Z-R, Pillai B-C, Lee CLR. Flash pyrolysis of jatropha oil cake in electrically heated fluidized bed reactor, Energy. 2010; 35: 2819–2823.

[21] Urban B, Shirazi Y, Maddi B, Viamajala S, Varanasi S. Flash pyrolysis of oleaginous biomass in a fluidized-bed reactor. Energy & Fuels 2017; 31(8): 8326-8334

[22] Papari S, Hawboldt K. A review on the pyrolysis of woody biomass to bio-oil: Focus on kinetic models. Renewable and Sustainable Energy Reviews, 2015; 52: 1580-1595.

[23] Makibar J, Fernandez-Akarregi A-R, Amutio M, Lopez G, Olazar M. Performance of a conical spouted bed pilot plant for bio-oil production by poplar flash pyrolysis. Fuel processing technology 2015; 137: 283-289.

[24] Oyebanji J-A, Okekunle P-O, Lasode O-A, Oyedepo S-O. Chemical composition of bio-oils produced by fast pyrolysis of two energy biomass. Biofuels 2018; 9(4): 479-487.

[25] Lazzari E, Schena T, Primaz C-T, da Silva Maciel G-P, Machado M-E, Cardoso C-A, Caramão E-B. Production and chromatographic characterization of bio-oil from the pyrolysis of mango seed waste. Industrial Crops and Products, 2016; 83: 529-536.
[26] Wang S, Gao B, Zimmerman A-R, Li Y, Ma L, Harris W-G, Migliaccio K-W. Physicochemical and sorptive properties of biochars derived from woody and herbaceous biomass. Chemosphere, 2015; 134: 257-262.
[27] Ahmad M, Rajapaksha A-U, Lim J-E, Zhang M, Bolan N, Mohan D. Biochar as a sorbent for contaminant management in soil and water: a review. Chemosphere 2014; 99:19–33
[28] Tag A-T, Duman G, Ucar S, Yanik J. Effects of feedstock type and pyrolysis temperature on potential applications of biochar. Journal of analytical and applied pyrolysis, 2016; 120: 200-206.
[29] Ábrego J, Plaza D, Luño F, Atienza-Martínez M, Gea G. Pyrolysis of cashew nutshells: Characterization of products and energy balance. Energy, 2018; 158: 72-80.
[30] Ma Z, Chen D, Gu J, Bao B, Zhang Q. Determination of pyrolysis characteristics and kinetics of palm kernel shell using TGA–FTIR and model-free integral methods. Energy Conversion and Management, 2015; 89: 251-259.
[31] Bensidhom G, Hassen-Trabelsi A-B, Alper K, Sghairoun M, Zaafouri K, Trabelsi I (2018). Pyrolysis of Date palm waste in a fixed-bed reactor: Characterization of pyrolytic products. Bioresource technology 2018; 247: 363-369.
[32] Quan C, Gao N, Song Q. Pyrolysis of biomass components in a TGA and a fixed-bed reactor: Thermochemical behaviors, kinetics, and product characterization. Journal of Analytical and Applied Pyrolysis 2016; 121: 84-92.
[33] Pattiya A, Suttibak S. Production of bio-oil via fast pyrolysis of agricultural residues from cassava plantations in a fluidised-bed reactor with a hot vapour filtration unit. Journal of Analytical and Applied Pyrolysis 2012; 95: 227-235.
[34] Xiu S, Shahbazi A. Bio-oil production and upgrading research: A review. Renewable and Sustainable Energy Reviews 2012; 16(7): 4406-4414.
[35] Farooq A, Shafaghat H, Jae J, Jung S-C, Park Y-K. Enhanced stability of bio-oil and diesel fuel emulsion using Span 80 and Tween 60 emulsifiers. Journal of environmental management 2019; 231, 694-700.
[36] Vispute T-P, Huber G-W. Production of hydrogen, alkanes and polyols by aqueous phase processing of wood-derived pyrolysis oils. Green Chemistry 2009; 11(9): 1433-1445.
[37] Fisk C, Crofcheck C, Crocker M, Andrews R, Storey J, Lewis Sr S. Novel approaches to catalytic upgrading of bio-oil. In 2006 ASAE Annual Meeting (p. 1). American Society of Agricultural and Biological Engineers 2006.
[38] Nokkosmäki M-I, Kuoppala E-T, Leppämäki E-A, Krause AOI. Catalytic conversion of biomass pyrolysis vapours with zinc oxide. Journal of Analytical and Applied Pyrolysis 2000; 55(1): 119-131.
[39] Wang J-J, Chang J, Fan J. Upgrading of bio-oil by catalytic esterification and determination of acid number for evaluating esterification degree. Energy & Fuels 2010; 24(5): 3251-3255.
[40] Zhang X, Wang T, Ma L, Zhang Q, Jiang T. Hydrotreatment of bio-oil over Ni-based catalyst. Bioresource technology 2013; 127: 306-311.
[41] Hew K-L, Tamidi A-M, Yusup S, Lee K-T, Ahmad M-M. Catalytic cracking of bio-oil to organic liquid product (OLP). Bioresource technology 2010; 101(22): 8855-8858.
[42] Padmaja K-V, Atheya N, Bhatnagar A-K, Singh K-K. Conversion of Calotropis procera biocrude to liquid fuels using thermal and catalytic cracking. Fuel 2009; 88(5): 780-785.
[43] Cheng D, Wang L, Shahbazi A, Xiu S, Zhang B. Catalytic cracking of crude bio-oil from glycerol-assisted liquefaction of swine manure. Energy Conversion and Management 2014; 87: 378-384.

[44] Kechagiopoulos P-N, Voutetakis S-S, Lemonidou A-A, Vasalos I-A. Hydrogen production via steam reforming of the aqueous phase of bio-oil in a fixed bed reactor. Energy & Fuels 2006; 20(5): 2155-2163.

[45] Garcia L, French R, Czernik S, Chornet E. Catalytic steam reforming of bio-oils for the production of hydrogen: effects of catalyst composition. Applied Catalysis A: General 2000; 201(2): 225-239.

[46] Forzatti P, Lietti L. Catalyst deactivation. Catal Today 1999; 52: 165–81.

[47] Yuan X, Ding X, Leng L, Li H, Shao J, Qian Y, Zeng G. Applications of bio-oil-based emulsions in a DI diesel engine: The effects of bio-oil compositions on engine performance and emissions. Energy 2018; 154: 110-118.

[48] Ramadhas A-S, Jayaraj S, Muraleedharan C. Characterization and effect of using rubber seed oil as fuel in the compression ignition engines. Renewable energy 2005; 30(5): 795-803.

[49] Van de Beld B, Holle E, Florijn J. The use of pyrolysis oil and pyrolysis oil derived fuels in diesel engines for CHP applications. Applied energy 2013; 102: 190-197.

[50] de Castro D-A-R, da Silva Ribeiro H-J, Ferreira C-C, de Andrade Cordeiro M, Guerreiro L-H-H, Pereira A-M, Oliveira R-L. Fractional Distillation of Bio-Oil Produced by Pyrolysis of Açaí (Euterpe oleracea) Seeds. In *Fractionation.* IntechOpen 2019..

[51] Buffi M, Cappelletti A, Rizzo AM, Martelli F, Chiaramonti D. Combustion of fast pyrolysis bio-oil and blends in a micro gas turbine. Biomass and bioenergy 2018; 115: 174-185.

[52] Kasper J-M, Jasas G-B, Trauth R-L. Use of pyrolysis-derived fuel in a gas turbine engine. In ASME 1983 International Gas Turbine Conference and Exhibit (1983, March) (pp. V003T06A016-V003T06A016). American Society of Mechanical Engineers 1983.

[53] Xu X, Cao X, Zhao L, Zhoua H, Luo Q. Interaction of organic and inorganic fractions of biochar with Pb(II) ion: further elucidation of mechanisms for Pb(II) removal by biochar, RSC Adv. 2014; 4: 44930–44937

[54] Ngan A, Jia C-Q, Tong S-T. Production, Characterization and Alternative Applications of Biochar. In Production of Materials from Sustainable Biomass Resources(pp. 117-151). Springer, Singapore 2019.

[55] Gupta M, Pramanik S, Ravikrishna R-V. Development of a syngas-fired catalytic combustion system for hybrid solar-thermal applications. Applied Thermal Engineering, 2016; 109: 1023-1030.

[56] Kan T, Strezov V, Evans T-J. Lignocellulosic biomass pyrolysis: A review of product properties and effects of pyrolysis parameters. Renewable and Sustainable Energy Reviews 2016; 57: 1126-1140.

[57] Shemfe M-B, Whittaker C, Gu S, Fidalgo B. Comparative evaluation of GHG emissions from the use of Miscanthus for bio-hydrocarbon production via fast pyrolysis and bio-oil upgrading. Applied Energy 2016; 176: 22-33.

[58] Hsu D-D. Life cycle assessment of gasoline and diesel production via pyrolysis and hyroprocessing. Biomass and Bioenergy2012: 41-47.

6

Torrefaction of Biomass

6.1 Introduction

Torrefaction is a thermophilic process of biomass conversion into a solid, friable, homogenous coal-like material, which has better fuel characteristics than the original biomass. Torrefaction involves the heating of biomass material in the absence of air at 250 to 300°C temperature. Torrefaction produces high-grade hydrophobic nature solid biofuels from various streams of woody biomass or agro residues.Torrefied biomass has high energy density then the original biomass which make torrefied biomass as renewable alternate source for power generation in coal power plant. Torrefied biomass being hydrophobic in nature can be stored easily for longer duration with negligible biological degradation. At 250-300°C temperature volatilization has been taking place and around 30% mass of the biomass is reduced as compared to original untorrefied biomass with net energy loss of approximately around 10%.The end product is a predictable, homogeneous, hydrophobic, high value solid biofuel with far higher energy density and calorific value than the original biomass feedstock. During the torrefaction process a combustible gas is released, which can also utilised to provide heat to the process. The process diagram of torrefaction process is shown in Fig. 6.1.

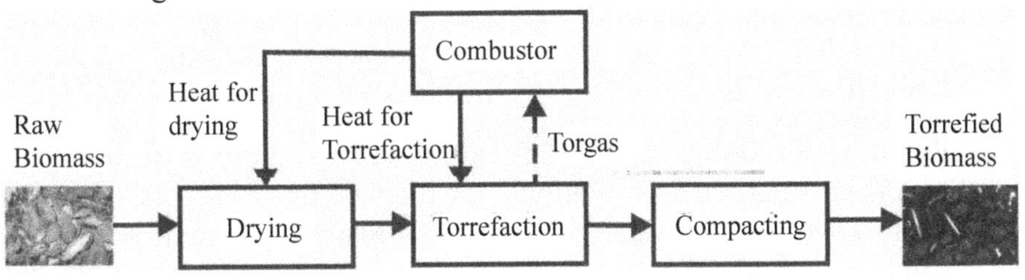

Fig. 6.1: Basic Torrefaction Process

6.2 Torrefaction Process

The total torrefaction process consists of drying, torrefaction and cooling. At beginning of the process the moisture in the biomass is released by drying.

When the temperature of biomass ranges between 200°C to 300°C, only in this temperature range, the torrefaction decomposition reactions occur and the duration of the torrefaction stage is around 30 minutes[1-4]. After torrefaction, the last stage is cooling the biomass from 200°C to the ambient temperature.

Bergman[5] classified the thermo-chemical changes in biomass during torrefaction in five different stages. as shown in Fig 6.2.

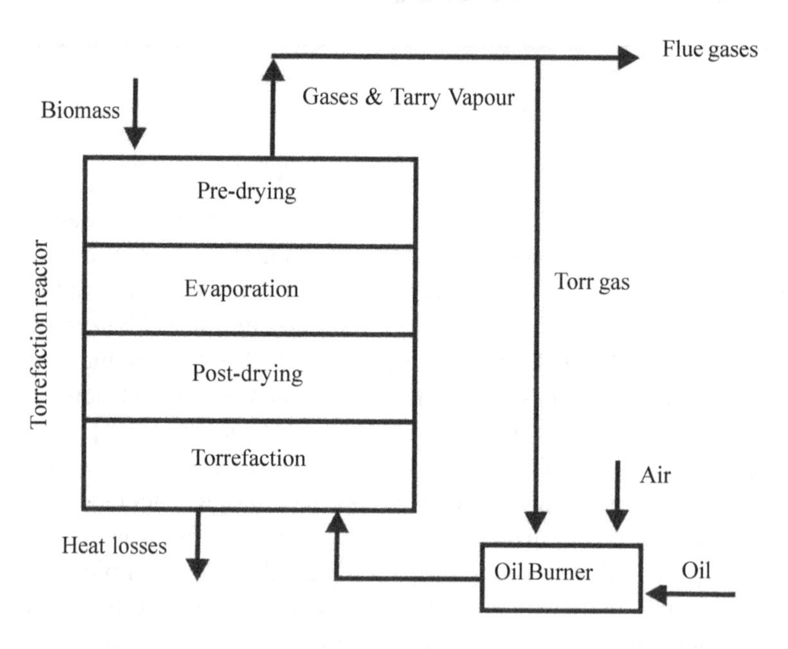

Fig. 6.2: Flow Chart of Torrefaction Process

1. Regime A (50-120°C): This is a nonreactive drying regime where there is a loss in physical moisture in biomass but no change in its chemical composition[6].
2. Regime B (120-150 °C): This regime is separated out only in case of lignin that undergoes softening, which make it serve as a binder.
3. Regime C (150-200 °C): This is called "reactive drying" regime that results in structural abnormality of the biomass that cannot be regained upon wetting. This produces shortened polymers that condense within solid structures[5].
4. Regime D (200-250 °C): This regime along with regime (E) constitutes torrefaction zone for hemicellulose. This regime is characterized by limited devolatilization and carbonization of solids structure formed in regime[6] (C).

5. Regime E (250-300 °C): This is the higher part of torrefaction process. Extensive decomposition of hemicellulose into volatiles and solid products takes place.

A typical mass and energy balance of torrefaction reveals that, 70% of the mass is retained as a solid product, containing 90% of the initial energy content. During torrefaction process about 30% of total mass is converted into gases, but containing only 10% of the energy content of the biomass. Hence, a considerable energy densification can be achieved[7], typically by a factor of 1.3 on mass basis. This example points out one of the fundamental advantages of the process, which is the high transition of the chemical energy from the feedstock to the torrefied product, whilst fuel properties are improved. This is in contrast to the classical pyrolysis process that is characterized by an energy yield of 55-65% in advanced concepts down to 20% in traditional ones[5].

Torrefaction is one of the techniques which have been studied recently to improve the grindability and energy density or heat value (HV) of biomass.Torrefaction mainly removes moisture from the biomass and causing it to become hydrophobic due to destruction of hydroxyl (OH) groups in the biomass allowing for easier storage and transportation[8]. The hydrophobic nature of torrefied biomass also results in less energy being required to process the biomass. The improved properties of torrefied biomass suggest that it is viable as a renewable energy source that is better suited for use in co-firing and gasification applications.

6.3 Design of Torrefaction Unit

6.3.1 Design Approach

The whole process of torrefaction can be divided into four stages (Heating zone, Drying zone, Post drying zone and Torrefaction zone) as shown in Fig. 6.3. For the sake of convenience, they were combined into four functional zones in a reactor:

1. Zone - A (Heating)
2. Zone - B (Drying)
3. Zone - C (Heating zone for torrefaction)
4. Zone - D (Torrefaction)

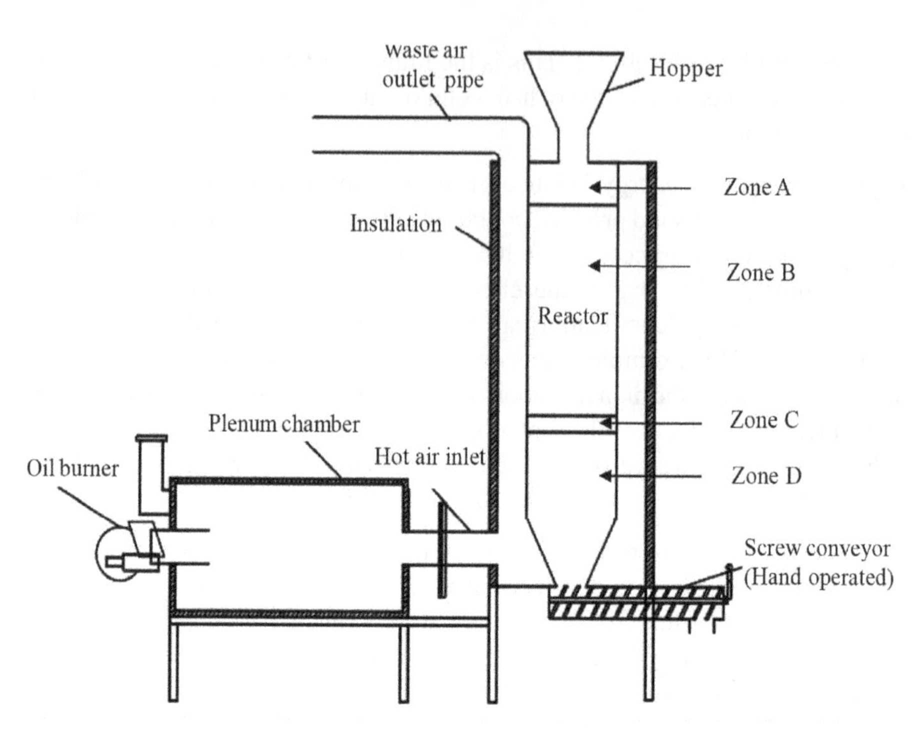

Fig. 6.3: Schematic of Torrefaction Unit

The assumption made for the design of torrefaction unit is presented in Table 6.1. These assumptionare commonaly available at local aras and gives best resent for torrefaction of biomass. Based on these assumptions further design was made and presented about

Table 6.1: Assumptions for Design of Biomass Torrefaction System

Sr. No.	Particulars	Specifications
1.	Product	Groundnut shell
2.	Capacity of dryer	50 kg/h
3.	Initial temperature	25°C
4.	Bulk density of material	150 kg/m^3
5.	Torrefied temperature	280°C
6.	Moisture content of material	10%
7.	Mass yield	70%
8.	Calorific value of material	15.70MJ/kg
9.	Calorific value of diesel	44.79MJ/kg
10.	Temperature hot gas at inlet	600°C
11.	Temperature of exist gas	105°C
12.	Air fuel ratio	14.6

6.3.2 Residence Time

Solid biomass moves under gravity with an acceptable space velocity provides the required residence time in different zones of reactor. This residence time was calculated using following equations for different zones given by Tumuluru[6].

i For Zone A

$$t = -Ln\left(\frac{\left(\dfrac{T - T_e}{T_o - T_e}\right)}{\left(\dfrac{hS}{\rho CV}\right)^{-1}}\right)$$

Where,

t = Residence time, (minute)

T = Temperature at time t, (°C)

T_e = Equilibrium temperature, (°C)

T_o = Initial temperature, (°C)

h = Average heat transfer coefficient, (W/m^2K)

s = Total surface area, (m^2)

ρ = Bulk density, (kg/m^3)

CV = Specific heat of material, (J/kgK)

$$t = \left(\frac{\left(\dfrac{100-160}{25-160}\right)}{\left(\dfrac{10 \times 0.00208}{150 \times 2000 \times \dfrac{0.0000032}{0.5}}\right)^{-1}}\right)$$

$$t = -Ln\left(\frac{0.44}{0.39}\right)$$

$$t = 2.69 \text{ min}$$

ii For Zone B

$$\left(\frac{M - M_e}{M_o - M_e}\right) = e^{-kt^n}$$

Where,

M = Moisture content at time t, (%)

Me = Equilibrium moisture content, (%)

Mo = Initial moisture content, (%)

k = Drying constant for the material

n = constant (exponent that improves the performance of the drying)

$$\left(\frac{0.1 - 0}{0.4 - 0}\right) = e^{-0.09 \times t^{1.15}}$$

t = 13.8 min

iii For Zone C

$$t = -Ln\left(\frac{\left(\frac{T - T_f}{T_o - T_f}\right)}{\left(\frac{hS}{\rho C_p V}\right)^{-1}}\right)$$

Where,

T = Temperature at time t, (°C)

= Equilibrium temperature at which the torrefaction reaction take place, (°C)

= Initial temperature, (°C)

= Average heat transfer coefficient, (W/m²K)

= Total surface area, (m²)

= Bulk density, (kg/m³)

= Specific heat of material, (J/kgK)

$$t = -Ln\left(\frac{\left(\frac{250-300}{150-300}\right)}{\left(\frac{15\times 0.00208}{150\times 2000\times \frac{0.0000032}{0.5}}\right)^{-1}}\right)$$

$$t = -Ln\left(\frac{0.33}{0.975}\right)$$

t=1.13 min

iv For Zone D

$$\left(\frac{M-M_e}{M_o-M_e}\right) = e^{-kt^n}$$

Where,

M = Mass concentrationof volatiles at time t, (%)

M_e = Equilibrium mass concentrationof volatiles, (%)

M_o = Initial mass concentrationof volatiles, (%)

k = Mass loss constant for the material

$$\left(\frac{0.1-0}{0.35-0}\right) = e^{-0.09\times t^{1.0}}$$

t = 7.78 min

6.3.3 Calculation of the Total Volume of Reactor

The volume of different zones was calculated to provide the specified solid residence time for the biomass feed and required yield using following formula given by Basu[9].

$$V = \frac{tW_t}{\delta_{bulk}}$$

Where,

t = Residence time, (minute)

W_t = Production rate, (kg/min)

δbulk = Bulk density of material

V = Volume of specified zone

$V_{total} = V_A + V_B + V_C + V_D$

Where,

V_A = Volume of zone A

V_B = Volume of zone B

V_C = Volume of zone C

V_D = Volume of zone D

$$V_B = \frac{tW_t}{\varrho_{bulk}}$$

$$V_B = \frac{13.8 \times 0.84}{150} = 0.07728\, m^3$$

$$V_C = \frac{tW_t}{\varrho_{bulk}}$$

$$V_C = \frac{1.13 \times 0.84}{150} = 0.006328\, m^3$$

$$V_D = \frac{tW_t}{\varrho_{bulk}}$$

$$V_D = \frac{7.78 \times 0.84}{150} = 0.0435\, m^3$$

$V_{total} = V_A + V_B + V_C + V_D$

$V_{total} = 0.0150 + 0.07728 + 0.006328 + 0.0435 = 0.142\ m^3$

6.3.4 Calculate the Total Height of Reactor

The height of different zone was calculated using following formula.

$$h = \frac{V}{\pi r^2}$$

Where,

h = Height of specified zone

V = Volume of specified zone

r = Radius of the reactor

$h_{total} = h_A + h_B + h_C + h_D$

Where,

h_A = Height of zone A

h_B = Height of zone B

h_C = Height of zone C

h_D = Height of zone D

$$h_A = \frac{V_A}{\pi r^2}$$

$$h_A = \frac{0.0150}{\pi \times 0.18^2} = 0.147\,m$$

$$h_B = \frac{V_B}{\pi r^2}$$

$$h_B = \frac{0.07728}{\pi \times 0.18^2} = 0.75\,m$$

$$h_C = \frac{V_C}{\pi r^2}$$

$$h_C = \frac{0.006328}{\pi \times 0.18^2} = 0.062\,m$$

$h_D = h_{D1} + h_{D2}$

$h_D = 0.32 + 0.2041 = 0.5241$

Where,

$V_D = 0.0435\ m^3$, $V_{D1} = 0.0335\ m^3$, $V_{D2} = 0.01\ m^3$

$$h_{D1} = \frac{V_{D1}}{\pi r^2}$$

$$h_{D1} = \frac{0.0335}{\pi \times 0.18^2} = 0.32\,m$$

$$h_{D2} = \frac{V_{D2}}{A}$$

$$h_{D2} = \frac{0.01}{\frac{1}{2}(0.36+0.12)\times h} = 0.2041m$$

$$h_{total} = h_A + h_B + h_C + h_D$$

$$h_{total} = 0.14 + 0.75 + 0.062 + 0.5241$$

$$= 1.4761$$

$$= 1.5 \text{ m}$$

6.3.5 Feed Rate of Biomass

$$W_f = \left[\frac{W_t}{MY_{daf}(1-M-A)}\right]$$

Where,

W_f = Feed rate of biomass, (kg/s)

W_t = Capacity of the unit, (kg/s)

MY_{daf} = Desired mass yield, (%)

M = Moisture content of biomass, (%)

A = Ash content of biomass, (%)

$$W_f = \left[\frac{0.014}{0.7(1-0.15)}\right]$$

$$= 0.024 \text{ kg/s}$$

$$= 86.4 \text{ kg/hr}$$

6.3.6 Total Energy Required for Torrefaction

$$Q_{ph} = W_f C_{pw} (100 - T_o)$$

Where,

Q_{ph} = Preheating feed from room temperature, (kW)

W_f = Feed rate of biomass, (kg/s)

C_{pw} = Specific heat of wet biomass, (kJ/kg°C)

T_o = Initial temperature, (°C)

$Q_{dry} = W_f ML$

Where,

Q_{dry} = Theoretical heat load of dryer, (kW)

W_f = Feed rate of biomass, (kg/s)

M = Moisture content of biomass, (%)

L = Latent heat of vaporization of water, (kJ/kg)

$Q_{pd} = W_d C_{pd} (T_t - 100)$

Where,

Q_{pd} = Theoretical heat load of post-drying heating, (kW)

W_d = Dry biomass feed rate, (kg/s)

C_{pd} = Specific heat of dried biomass, (kJ/kg°C)

T_t = Torrefaction temperature, (°C)

$Q_{total} = Q_{ph} + Q_{dry} + Q_{pd}$

Where,

Q_total = Theoretical heat load of torrefier, (kW)

$Q_{ph} = 0.024 \times 1.46 (100 - 25) = 2.628$ kW

$Q_{dry} = 0.024 \times 0.15 \times 2260 = 8.1$ kW

$Q_{pd} = 0.0204 \times 0.269 (280 - 100) = 0.98$ kW

$Q_{total} = Q_{ph} + Q_{dry} + Q_{pd}$

$Q_{total} = 2.628 + 8.1 + 0.98 = 11.708 = 12$ kW

6.3.7 Amount of Air Required for Torrefaction

$$W_g = \frac{\left(W_v C_v + W_{vl} C_g\right) T_{gO} + W_t C_d T_t - W_f C_b T_O + W_v L}{C_g (T_{gi} - T_{gO})}$$

Where,

W_g = Flow rate of diluted hot gas entering the torrefier, (kg/s)

W_v = Weight of volatiles matter, (kg/s)

C_v = Specific heat of Steam, (kJ/kg°C)

W_{vl} = flow rate of volatile in fluid product leaving torrefier, (kg/s)

C_g = Specific heat of fuel gas, (kJ/kg°C)

T_{go} = Temperature of gas at exit of torrefier plant, (°C)

W_t = Production rate of torrefied biomass, (kg/s)

C_d = Specific heat of dry or torrefied biomass, (kJ/kg°C)

T_t = Torrefaction temperature, (°C)

W_f = Feed rate of biomass, (kg/s)

C_b = Specific heat of raw biomass, (kJ/kg°C)

L = Latent heat of vaporization of water, (kJ/kg)

T_{gi} = Temperature of gas at inlet of torrefier plant, (°C)

$$W_g = \frac{\begin{matrix}(0.0036\times1.89+0.0064\times1.13)105+0.014\times0.269\times280 \\ -0.024\times1.46\times25+0.0036\times2260\end{matrix}}{1.13(600-105)}$$

$$W_g = \frac{9.774}{559.35} = 0.01747\, kg/s$$

6.3.8 Amount of Fuel Required for Oil Burner

$$W_{oil} = \frac{1}{K-P}\times\frac{W_g}{W'_g}\left[\frac{C_g T_{gi}}{C_g T_{gO} + VL'_{fr}\ LHV_{vl}}\right]$$

Where,

W_{oil} = Flow rate of oil burnt, (kg/s)

K = Constant Parameter

P = Constant Parameter

W_g = Flow rate of diluted hot gas entering the torrefier, (kg/s)

W^i_g = Flow rate of fuel gases carrying with it moisture and product gases from the torrefaction, (kg/s)

C_g = Specific heat of fuel gas, (kJ/kg°C)

T_{gi} = Temperature of gas at inlet of torrefier plant, (°C)

T_{go} = Temperature of gas at exit of torrefier plant, (°C)

VL'_{fr} = Fraction of volatiles in the product gas of torrefaction

$$W_{oil} = \frac{1}{3948.69 - 674.19} \times \frac{0.01747}{0.02747} \left[\frac{1.13 \times 600}{1.13 \times 105 + 0.2329 \times 1286 \times 0.95} \right]$$

W_{oil} = 0.000132 kg/s

6.4 Development of Biomass Torrefaction System

The material used for the fabrication of torrefaction system are presented in Table 6.2.

Table 6.2: Material Used During Fabrication of Torrefaction System

Reactor chamber	Material
Inner cylinder	MS sheet, 1.5 mm
Outer cylinder	HRC (hot rolled sheet), 2.0 mm
Insulation	Glass wool
Outermost cover over insulation	Galvanized iron sheet, 0.5 mm
Stand	L shape MS angle
Plenum chamber	
Chamber cylinder	HRC sheet, 2.0 mm
Insulation	Glass wool
Outermost cover over insulation	Galvanized iron sheet, 0.5 mm
Hand operated screw conveyer	
Outer body	10 inch MS pipe
Screw	HRC sheet, 2.0 mm
Inner rod	1.5 inchØ MS rod
Handle	MS flat plate and pipe

The schematic of developed torrefaction system are illustrated in Fig. 6.4a, Fig. 6.4b, and Fig. 6.4c. The technical specifications of designed torrefaction system is presented in Table 6.3

Table 6.3: Technical Specifications of Biomass Torrefaction System (All dimensions in meter)
Reactor chamber

Zone	Volume (m^3)	Height (m)
Zone A	0.0150	0.147
Zone B	0.7728	0.75
Zone C	0.006328	0.062
Zone D	0.0435	0.5241
Total	0.142	1.5
Inner reactor cylinder (diameter)	0.36	
Outer reactor chamber (diameter)	0.60	

(Contd.)

Insulation over outer cylinder (thickness)	0.03
Total diameter of reactor	0.66
Inlet and outlet diameter	0.12
Plenum chamber	
Diameter of chamber	0.45
Insulation over chamber (thickness)	0.025
Length of chamber	0.92
Total diameter of plenum chamber	0.5
Inlet and outlet diameter	0.12
Hand operated of screw conveyer	
Length of screw conveyer	0.92
Diameter of screw conveyer	0.12
Required Oil burner Specification	
Amount of air required for torrefaction (kg/min)	1.0482
Theoretical oil consumption rate (kg/min)	0.00792
Theoretical feed rate of biomass (kg/min)	1.44
Theoretical total energy required for torrefaction (kW)	12

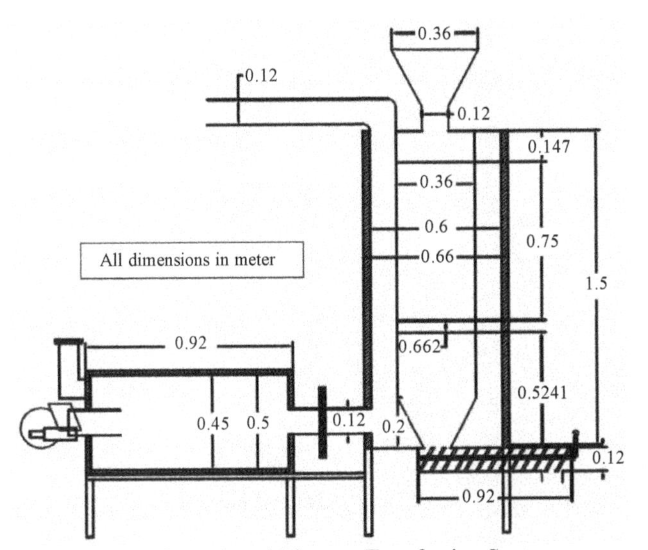

Fig. 6.4a: Dimensions of Developed Biomass Torrefaction System

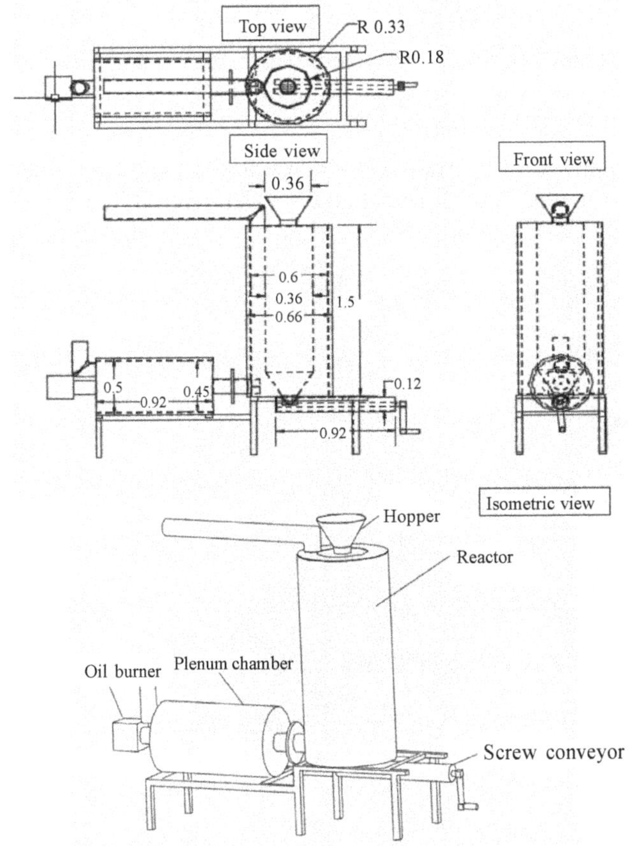

Fig. 6.4b: Details of Developed Biomass Torrefaction System

Fig. 6.4c: Developed Biomass Torrefaction System

6.5 Thermal Performance of Biomass Torrefaction System

Thermal performance of designed system was assessed under both no load and full load conditions. Groundnut shells were used as feed material during whole operation and industrial oil burner was used to supply process heat for external heating of the biomass. The K-type thermocouples were placed at different positions to record the temperature of different zones as shown in Fig 6.5.

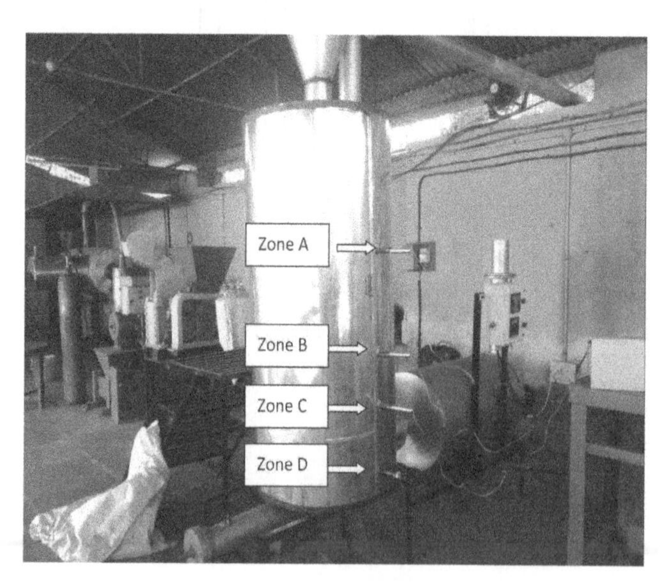

Fig. 6.5: Position of temperature sensors

6.5.1 No Load Performance

No load performance test was conducted to know the temperature profile along the different zone of torrefied reactor. The diesel burner was used for external heating to the biomass at desired temperature in the torrefaction system. The no load test was conducted for one hour and thermocouples were placed on different zone like zone A, zone B, zone C and zone D, respectively.

Fig. 6.6 reveals the temperature profile across the four zones. The maximum temperature for zone D, zone C, zone B and zone A were recorded about 281°C, 256°C, 232°C and 218°C respectively. While the average temperatures for zone D, zone C, zone B and zone A were found about 220°C, 183°C, 170°C and 164°C respectively, The average temperature for whole reactor chamber was found 192°C. Temperature data of no load performance is given in Appendix-B. during no test, burner consume about 0.32 liter of diesel oil. Based on no load temperature profile it is concluded that the developed system is suitable for torrefaction of agricultural crop residue.

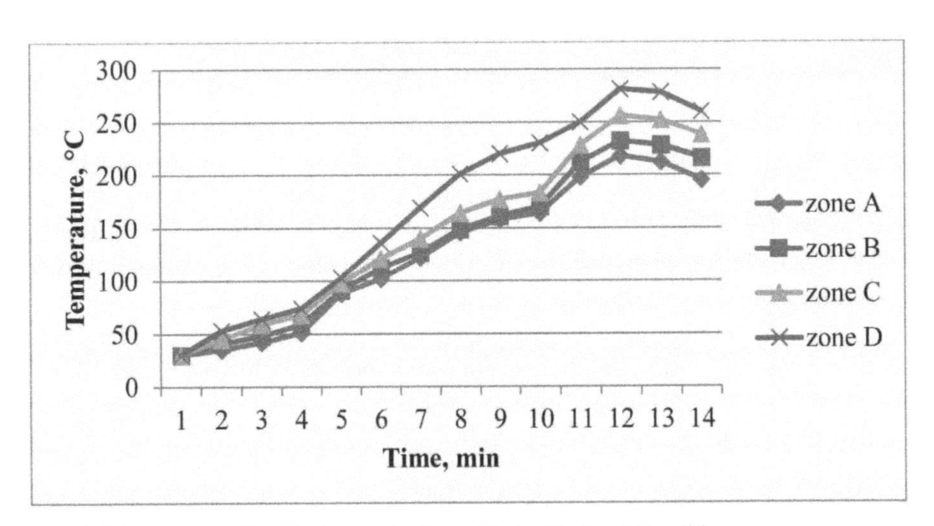

Fig. 6.6: Temperature Profile Across the Zones Under No Load Conditions

6.5.2 Full Load Performance

Full load performance of system was evaluated by loading the reactor with material to be torrified i.e. groundnut shells as shown in Fig 6.7. Torrefaction unit was tested initially for groundnut shells and it was procured from local oil industries. Multiple replications were carried out to assess the performance of the developed torrefaction system under full load condition. The results of three replications are presented under full load condition. Initially, torrefaction reactor was preheated in the range of 200°C, 250°C and 280°C temperature for three replications. About 10 kg groundnut shells having moisture content of 9.5 per cent were fed to the reactor for torrefaction. The raw groundnut shells were indirectly heated continuously for 10 minutes duration and the temperatures of the four zones in the reactor were recorded at 1 minute of time interval.

Fig. 6.7: Groundnut Shells

Fig 6.8 reveals temperature profile of first trial across the four zones where reactor was preheated initially up to 200°C temperature. The maximum temperature for zone D, zone C, zone B and zone A were recorded about 283°C, 276°C, 268°C and 248°C respectively. While the average temperatures for zone D, zone C, zone B and zone A were found about 230°C, 211°C, 198°C and 179°C respectively. The average temperature for whole reactor chamber was found 204°C. In this trial, burner consume about 0.35 liter of diesel oil.

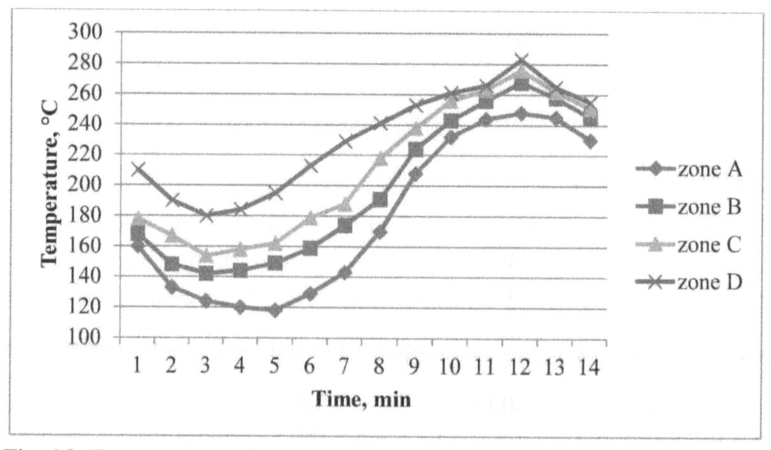

Fig. 6.8: Temperature Profile Across the Zones Under Full Load Condition (Trial 1)

Fig 6.9 reveals temperature profile of second trial across the four zones where reactor was preheated initially up to 250°C temperature. The maximum temperature for zone D, zone C, zone B and zone A were recorded about 314°C, 292°C, 279°C and 261°C respectively. While the average temperatures for zone D, zone C, zone B and zone A were found about 250°C, 236°C, 223°C and 209°C respectively, The average temperature for whole reactor chamber was 230°C and during second trial, burner consume about 0.4 liter of diesel oil.

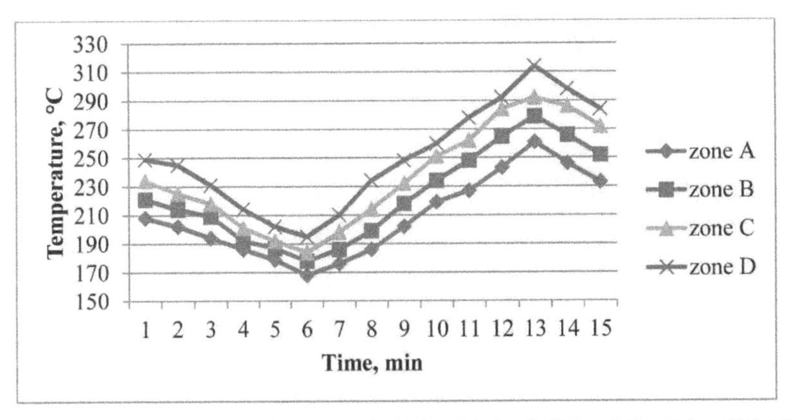

Fig. 6.9: Temperature Profile Across the Zones Under Full Load Condition (Trial 2)

Fig 6.10 reveals temperature profile of third trial across the zones where reactor was preheated initially up to 280°C temperature. The maximum temperature for zone D, zone C, zone B and zone A were recorded about 321°C, 312°C, 298°C and 279°C respectively. While the average temperatures for zone D, zone C, zone B and zone A were found about 275°C, 264°C, 253°C and 240°C respectively, The average temperature for whole reactor chamber was 258°C and during third trial, burner consume about 0.48 liter of diesel oil.

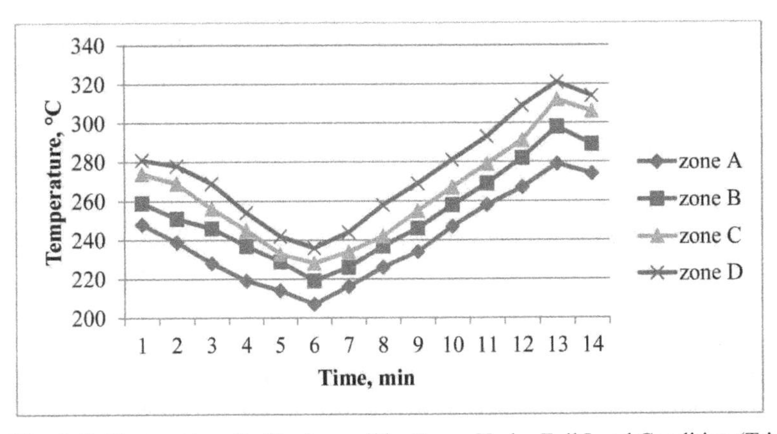

Fig. 6.10: Temperature Profile Across The Zones Under Full Load Condition (Trial 3)

Then, the material was removed from the reactor with the help of hand operated screw as shown in Fig 6.11 and its mass in kg and higher heating value in MJ/kg was assessed in laboratory.

Fig. 6.11: Working of Hand Operated Screw Conveyor

The results obtained from full load performance of the designed the torrefaction system with groundnut shells were evaluated after checking following parameters which are depicted in Table 6.4.

Table 6.4: Thermal Analysis of Developed Torrefication System

Thermal performance of system	Output		
Parameter	Trial 1	Trial 2	Trial 3
Average temperature, °C	204	230	258
Residence time, min	10	10	10
Initial mass fed in reactor, kg	10	10	10
Torrified product, kg	7.15	6.73	5.34
HHV of raw material fed, MJ/kg	15.70	15.70	15.70
HHV of torrified product, MJ/kg	17.31	20.66	22.76
Mass yield, per cent	71.5	67.30	53.4
Energy yield, per cent	78.84	88.56	77.4
Energy gain by torrified product per kg in per cent	10.26	31.6	44.9
Diesel consumption, L	0.35	0.4	0.48
HHV of diesel, MJ/kg	44.79	44.79	44.79
Energy required for torrefaction (MJ/batch)	1.57	1.79	2.15
Net energy (MJ/kg product)	15.74	18.87	20.61

From Table 6.4, the torrified product obtained in first replication had the higher heating value 17.31MJ/kg and mass yield 71.5 per cent. Thus, the energy yield of the product was 78.84 per cent in first replication. The torrified product obtained in second replication had the higher heating value 20.66MJ/kg and mass yield 6.73 per cent. Thus, the energy yield of the product was 88.56 per cent in second replication. In the similar fashion torrefied materials obtained during third replication had the higher heating value 22.76MJ/kg and mass yield 53.4 per cent. Thus, the energy yield of the product was 77.4 per cent in third replication.

Hence, it is clear that the energy yield of the torrified product was higher in second replication as compared to other trials. Thus, the second replication was found best amongst the three replications. Also, it is observed from the results that 10 minute time was sufficient for torrefaction of groundnut shells at average reactor temperature of 230°C. If the temperature exceeded than the average reactor temperature 230°C or if time exceeded than 10 minutes duration, carbonization of biomass was started. The Torrefied biomass obtained from developed torrefaction system and its briquettes are shown in Fig 6.12.

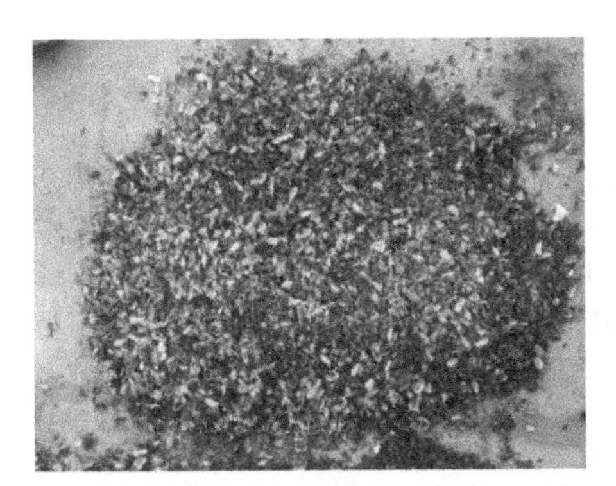

Fig. 6.12: Torrified Biomass and Its Briquettes

Example 6.1: A batch of 10 kg groundnut shell having higher heating value about 15.70 MJ/kg, torrefied at 258 °C and resulting 5.37 kg of torrefied biomass with higher heating value about 22.76 MJ/kg. The energy required during torrefaction of a batch is about 2.15 MJ. Calculate the energy density for this torrefaction process.

Solution

$$\textit{Energy required to torrefied a batch} = 2.15MJ$$

$$Energy\,retains\,in\,biomass = 22.76\frac{MJ}{kg}\times 5.34\,kg = 121.54\,MJ$$

Energy lost during torrefaction process

= HHV of Raw biomass – Energy retains in biomass

– Energy required for torrefaction

= 157 MJ – 127.54MJ – 2.15MJ

= 33.31 MJ

$$Energy\,lost = \frac{33.31}{157} \times 100$$

= 35.44%

Amount of energy conserve = 100 – 35.44

= 64.56%

$$Energy\,Densification = \frac{Energy\,conserve\ \%}{Mass\ yield\ \%}$$

$$= \frac{0.6456}{0.534} = 1.20$$

Example 6.2: One kilogram of groundnut shell having higher heating value about 15.70 MJ/kg torrefied at 258 °C. Solid content during torrefaction process was found about 0.534 kg with higher heating value of 22.54 MJ/kg. Approximately 0.15 kg of liquid was collected with 20 % moisture. Liquid collected during torrefaction process having heating value of 22 MJ/kg. Determine the energy content in gaseous part. Assume no losses.

Solution

$$Energy\,contained\,in\,torrefied\,biomass = 0.534\,MJ \times 22.54\frac{MJ}{kg}$$

= 12.04 MJ

Energy Contained in liquid = Liquid collected × organic portion × heating value

= 0.15 × 0.80 × 22.00

= 2.64 MJ

Energy balance = Heating value of raw biomass

= Energy in torrefied biomass + Energy liquid + Energy in gas

15.70 = 12.04 + 2.64 + energy in gas

Energy in gas = 1.02 MJ

Energy banace = Solid 76.68% + liquid 16.82% + Gas 6.50%

6.6 Mass and Energy Yield

Mass yield can be defined as it is ration of mass of torrefied biomass to mass of raw/ un torrefied biomass

$$Mass\,Yield = \frac{Mass\,of\,torrefied\,biomass}{Mass\,of\,raw\,biomass}$$

Similarly, energy yield is the ration of energy in torreifed biomass to energy in raw biomass

$$Energy\,Yield = \frac{Energy\,in\,torrefied\,biomass}{Energy\,in\,raw\,biomass}$$

6.7 Torrefaction Technologies

Many companies in the worldwide have developed torrefied technology for biomass upgradation. Table 6.5 provides details of different torrefaction reactor technologies and companies that have developed them. Large scale torrefied biomass production as shown in Fig.6.13 can be done using any of the following torrefaction reactor design.

Table 6.5: Torrefaction Reactor Technologies and Companies[10]

Torrefaction technologies, (based on reactor design)	Company (Developer)
Fixed Bed Reactor	Parker Autoclave Engineers (US)New Earth Eco Technologies (US)
Rotary drum reactor	CDS (UK), Torr-coal (NL), BIO3D (FR), EBES AG (AT), 4Energy Invest(BE), BioEndev/EPTS (SWE), Atmosclear S.A.(CH).
Screw type reactor	Picheney Rotary, BTG (NL), Biolake(NL), Foxcoal(NL), Agritech (US)
Multiple Hearth Furnace	CMI-NESA(BE), Wyssmont (US) Multiple plate, Integro Earth Fuels LLC (US).
Torbed reactor (rotating fluidized bed)	Toppel (NL) Entrained, Torftech group(UK).
Microwave reactor	Rotawave limited (UK), Can Biocoal (UK), Airex (CAN), Torrefaction Systems (US)
Compact moving bed reactor	ECN(NL), Thermya (FR), Buhler (GER)
Belt dryer	Strampoy Green Investment (NL), New Earth Eco Technology (US), Agri Tech Producers LLC RTF (US), 4EnergyInvest-AmelBiocoal(BE).

Key: AT- Austria, BE-Belgium, CA- Canada, CH-Switzerland, FR-France, IT-Italy, NL-Netherlands, GER-Germany, S-Spain, SWE-Sweden, UK-United Kingdom, US-United States of America.

Fig. 6.13: Torrefied Biomass Briquettes

Picture courtesy: The International Biomass Torrefaction Council (IBTC), Brussels, Belgium

6.8 Pros and Cons of Biomass Torrefaction

Torrefaction of biomass improves the physical characteristics of biomass, and thus the overall economics of the biomass utilization process for energy production.Torrefaction of biomass results in a high energy value solid homogenous biofuel which can be used as an alternate source of renewable fuel in coal power plant for power generation. The comparison of torrefied biomass briquettes with coal and normal wood pellets has been made in Table 2 whereas benefits of torrefied biomass briquettes over coal and normal biomass is highlighted below in following section.

6.9 Benefits of Torrefied Biomass Briquettes over Coal

- Grinds & burns like coal – existing infrastructure can be used
- Lower feedstock costs
- Lower shipping costs
- Minimal de-rating of the power plant
- Provides non-intermittent renewable energy
- Lower sulphur and ash content (compared with coal)

6.10 Benefits of Torrefied Biomass Briquettesover Normal Biomass

- Higher calorific value
- More homogeneous product
- Hydrophobic nature/water repellent:
    - Transport and material handling is less expensive & easier
    - Outdoor storage possible
    - Less expensive storage option
    - Significant loss of energy due to re-absorption of moisture in biomass (pellets) is saved
- Negligible biological activity (resistance to decomposition)
    - Longer storage life without fuel degradation
- Low O/C ratio
    - higher yield during gasification
- Higher bulk density
- Excellent grindability
- Higher durability
- Smoke producing compounds removed

6.11 Drawbacks of Biomass Torrefaction

- Some of the energy content (around 10 %) in original biomass is loosed due to torrefaction process
- Limited knowledge of torrefcation process temperature, properties of torrefied biomass and composition of volatiles released during the process.
- Torrefcation technology is not yet commercially implemented anywhere in India.

6.12 Utility/Application/Market of Torrefied Biomass

- Torrefied biomass pellets can be used as source of fuel in improved cookstove.
- Torrefied biomass briquettes can be used in boiler instead of coal as a source of fuel.

- Torrefied biomass can be used as an alternate source of feedstock in replace of coal for power generation.
- It can be used as source of feedstock to generated transportation fuel by Fischer- Tropsch process.
- Torrefied biomass can also be utilized for heating of houses in cold seasons.

References

[1] Rousset P, MacEdo L, Commandré J M, Moreira A. Biomass torrefaction under different oxygen concentrations and its effect on the composition of the solid by-product. Journal of Analytical and Applied Pyrolysis 2012: 96:86–91.

[2] Bach QV, Skreiberg O. Upgrading biomass fuels via wet torrefaction: A review and comparison with dry torrefaction. Renewable and Sustainable Energy Reviews 2016; 54: 665–677.

[3] Brue J, Brue JD. Development of self-selection torrefaction system. Master's thesis. Iowa state university, Ames, Iowa 2012.

[4] Dudgeon R, Charlotte N. An aspen plus model of biomass torrefaction. University Turbine Systems Research (UTSR) Fellowship, Electric Power Research Institute (EPRI) Charlotte, NC2009.

[5] Bergman PC, Boersma R, Zwart RWR, Kiel JH. Torrefaction for biomass co-firing in existing coal-fired power stations. Energy Research Centre of the Netherlands ECN ECNC05013, (July), 2005:71.

[6] Tumuluru JS, Sokhansanj S, Wright CT, Boardman RD. Biomass Torrefaction Process Review and Moving Bed Torrefaction System Model Development. Idaho National Laboratory Biofuels and Renewable Energy Technologies Department Idaho Falls, Idaho, INL/EXT-10-19569 Rev. . 2010;1: 1-44.

[7] Peng JH, Bi XT, Sokhansanj S, Lim CJ. Torrefaction and densification of different species of softwood residues. Fuel 2013; 111, 411-421.

[8] Dhungana A, Dutta A, Basu P. Torrefaction of non-lignocellulose biomass waste. Canadian Journal of Chemical Engineering 2012; 90(1): 186–195.

[9] Basu P. Biomass gasification and pyrolysis: practical design and theory. Academic press 2010.

[10] Eseyin AE, Steele PH, Pitman CU. Torrefaction Trends. Bioresources, 2015; 10(4): 8812-8858.

7

Biomass Gasification Technology

7.1 Introduction

Biomass gasification is age old technology originated in 1800 century for lighting and cooking purposes (Sansaniwal et al., 2017). Biomass typically contents biopolymers such as cellulose, hemicellulose and lignin in addition to the presence of average composition of $C_6H_{10}O_5$ which depends upon physical characteristics of biomass[1]. Oxygen is required for combustion of every fuel. For complete combustion of biomass stoichiometric air to fuel ratio required varies from 6:1 to 6.5:1 with CO_2 and H_2O as end product.

Biomass gasification is an efficient and environmentally friendly way to produce energy[2]. Gasification process is nothing but it is a conversion of solid fuel into gaseous fuel for wide applications. This whole process completed at elevated temperature range of 800-1300 °C with series of chemical reaction that is why it come under thermo chemical conversion[3]. Thermo chemical processes are most commonly employed for converting biomass into higher heating value fuels[4]. Major thermal conversion route is include direct combustion to provide heat, liquid fuel and other elements to generate process heat for thermal and electricity generation is summaries in Fig. 7.1.

Biomass as a feedstock is more promising than coal for gasification due to its low sulfur content and less reactive character. The biomass fuels are suitable for the highly energy efficient power generation cycles based on gasification technology. It is also found suitable for cogeneration. The combustion in gasifier take place in limited supply of oxygen it may be called partial combustion of solid fuel[5]. The resulting gaseous product called producer gas is an energy rich mixture of combustible gas H_2, CO, CH_4 and other impurities such as CO_2, nitrogen, sulfur, alkali compounds and tars[6].

In gasification process incomplete combustion of biomass with less oxygen supply is required with stoichiometric air to fuel ration of 2:1 to 2.3:1. Biomass gasification occurs through a sequence of complex thermochemical reactions in different zones. The design of biomass gasifier depends on fuel availability, shape and size, moisture content, ash content and end user applications. Different types of biomass gasifiers are available depending upon various sizes and design as per requirement which mainly classified in to fixed bed and fluidized bed type of gasifiers[7]. Fixed bed gasifiers are further classified in to updraft, downdraft and cross draft type of gasifiers according to the way of interaction of either air/oxygen or steam with biomass.

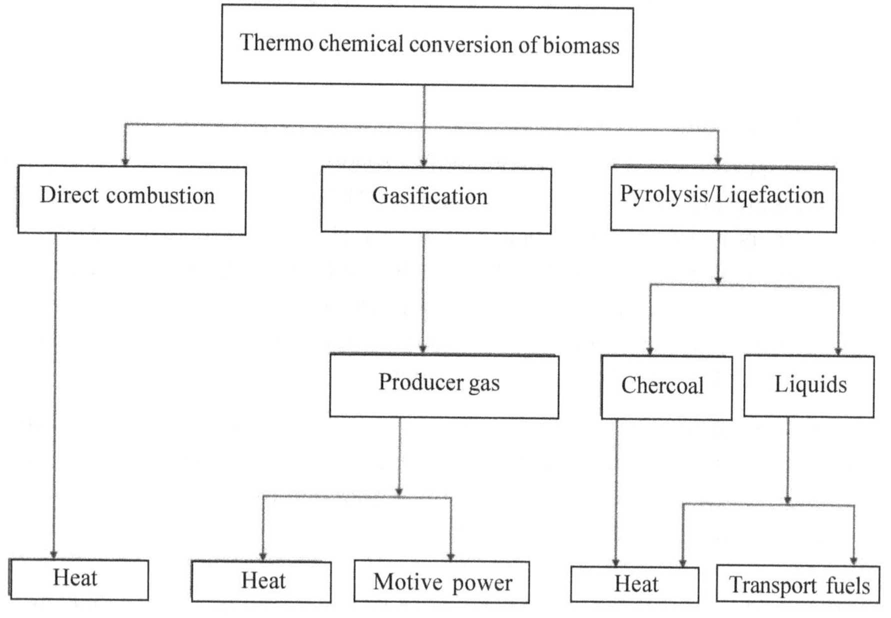

Fig. 7.1: Thermo Chemical Conversion Route of Biomass

7.2 History and Development

The history of gasification dates back to seventeenth century. Since the conception of idea, gasification has passed through several phases of development. Yearwise development of the technology is given below[8].

1669 : Thomas Shirley conducted crude experiments with carborated hydrogen

1699 : Dean Clayton obtained coal gas from pyrolitic experiment

1788 : Robert Gardner obtained the first patent with regard to gasification

1792 : First confirmed use of producer gas reported, Murdoc used the gas generated from coal to light a room in his house. Since then, for many years coal gas was used for cooking and heating

1801 : Lampodium proved the possibility of using waste gases escaping from charring of wood

1804 : Fourcroy found the water gas by reaction of water with a hot carbon

1812 : developed first gas producer which uses oil as fuel

1840 : First commercially used gasifier was built in France

1861 : Real breakthrough in technology with introduction of Siemens gasifier. This gasifier is considered to be first successful unit

1878 : Gasifiers were successfully used with engines for power generation

1900 : First 600 hp gasifier was exhibited in Paris. Thereafter, larger engines upto 5400 hp were put into service

1901 : J.W. Parker run a passenger vehicle with producer gas

After 1901 : In the period 1901-1920, many gasifier-engine systems were sold and used for power and electricity generation

1930 : Nazi Germany accelerated effort to convert existing vehicles to producer gas drive as part of plan for national security and independence from imported oil

1939 : About 2,50,000 vehicles were registered in the Sweden. Out of them, 90 % were converted to producer gas drive. Almost all of the 20,000 tractors were operated on producer gas. 40 % of the fuel used was wood and remainder charcoal

After 1945 : After end of second world war, with plentiful gasoline and diesel available at cheap cost, gasification technology lost glory and importance

1950 1970 : During these decades, gasification was "Forgotten Technology". Many governments in Europe to felt that consumption of wood at the prevailing rate will reduce the forest, creating several environmental problems

After 1970 : The year 1970's brought a renewed interest in the technology for power generation at small scale. Since then work is also concentrated to use fuels other than wood and charcoal.

7.3 Classification of Biomass Gasifiers

Design of gasifier depends upon type of fuel used, air introduction in the fuel column and type of combustion bed as shown in Fig.7.2.

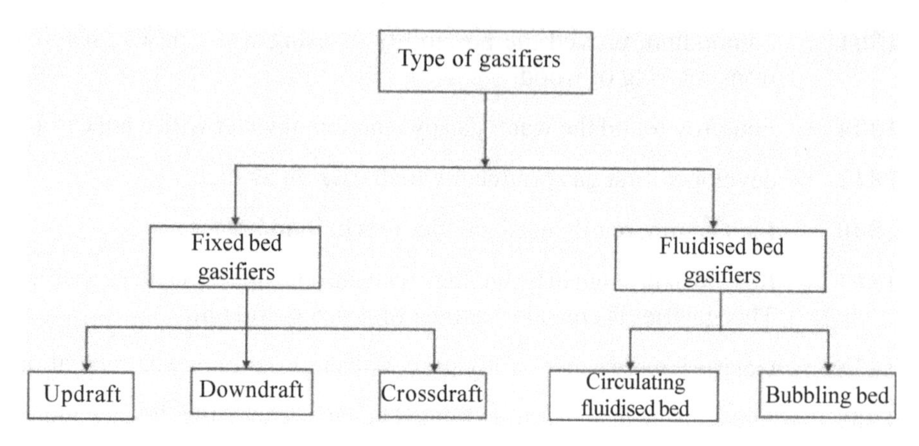

Fig.7.2: Classification of Gasifiers

7.3.1 Fixed Bed Gasifier

Fixed bed gasifiers have a fixed grate to support the biomass fuel fed in the reactor for gasification. Design and operation of these gasifiers are relatively easy as compare to fluidized bed reactor. These are basically classified according to the movement of air or steam in the reactor in to updraft, downdraft and crossdraft type of gasifier. The fixed types of gasifiers are generally preferred for small to medium scale applications and in this case erosion of the reactor body is less[9].

7.3.1.1 Updraft Gasifier

In updraft gasifier gasifying agent such as air or steam are introduced at the bottom section of gasifier to interact with biomass which is fed from the top of reactor (Fig.7.3). The generated producer gases after gasification exits from the top side of the gasifier hence this gasifier is also called as counter current type of gasifier. After biomass gasification high calorific value syngas as end product exits from the top side of the reactor. This type of gasifier has highest thermal efficiency as the generated hot gases passed through biomass fuel bed which left the gasifier unit at low temperature whereas some part of sensible heat of generated producer gases is used for drying of biomass fuel (Malik and Mahapatra, 2013). Apart from high thermal efficiency updraft gasifiers have some other main advantages such as small pressure drop and slight tendency of slag formation. These gasifiers are suitable for the applications where the high flame temperature is required. However updraft gasifiers has some drawbacks such as sensitivity to tar and moisture content of the biomass, low gas production, long start up time of the engine and poor reaction capability of the system[7].

7.3.1.2 Downdraft Gasifier

In downdraft gasifiers air or steam interacts with the solid biomass fuel in the downward direction therefore generated producer gases flow downward in the co-current direction and exits from the bottom side of the reactor. Downdraft type of gasifier as shown in Fig.7.4 is also called as co-current type of biomass gasifier. The end products of the pyrolysis and drying zone are forced to pass through oxidation zone of the reactor for thermal cracking which yield less tar content in the final product and hence better quality of fuel. This gasifier is suitable for small scale decentralize power generation because of low tar and particulate matter content in the finally produced syngas[10].

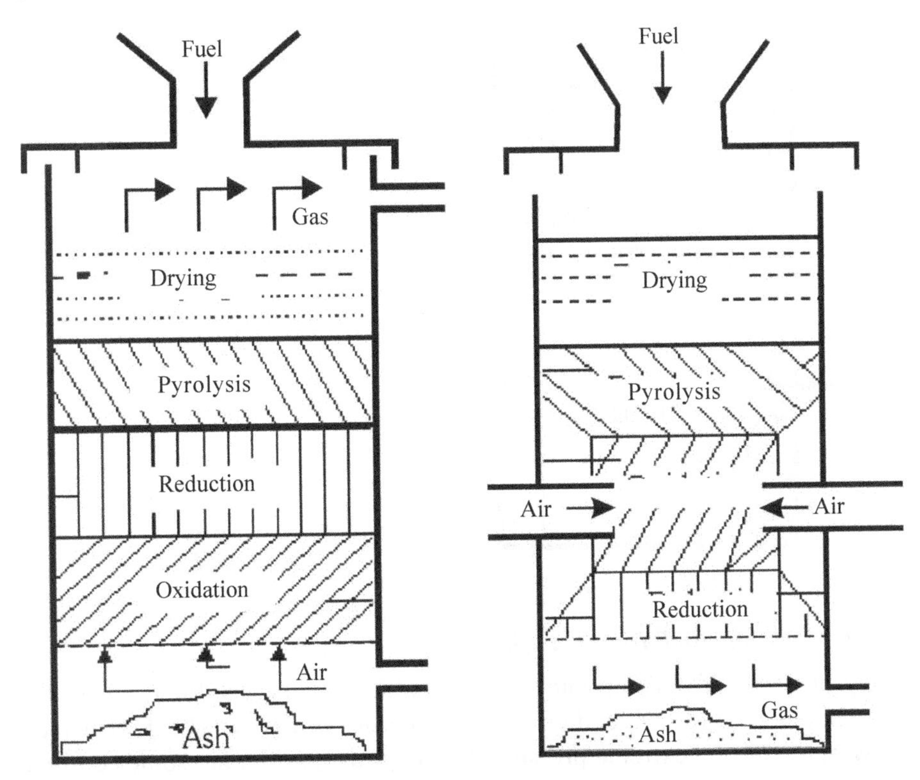

Fig. 7.3: Updraft Gasifier

Fig. 7.4: Downdraft Gasifier

7.3.1.2 Crossdraft Gasifier

Crossdraft gasifier as shown in Fig. 7.5 is one of the simplest types of gasifier design in which biomass fuel enters from the top of the reactor whereas air is provided from the side of the reactor instead of top or bottom side of the reactor. Crossdraft gasifier has certain advantages as compare to updraft and downdraft type of gasifiers but it is not of ideal type. The Crossdraft gasifier has separate zones of ash bin, fire and reduction which limit the biomass fuel used for operation with less ash content in the final produced syngas[11]. Crossdraft gasifier has certain advantages such as fast response against biomass fed in the reactor, small start time, compatible with dry air blast, flexible syngas production, and comparatively shorter design height. But apart from some major advantages crossdraft gasifier has certain drawbacks such as incapability to handle high tar content and very small biomass particles. This gasifier produces high temperature syngas with poor reduction rate of carbon dioxide. Therefore, Crossdraft gasifier has limited application in the field and not much work has been reported in the literature[7].

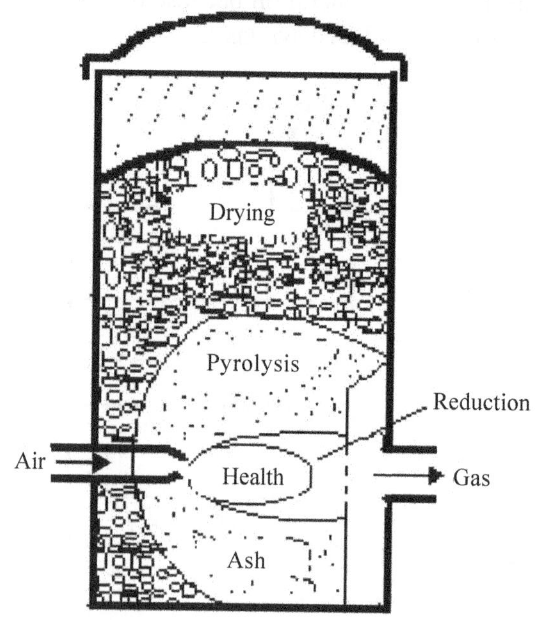

Fig. 5: Cross Draft Gasifier

7.3.2 Fluidized Bed Type Gasifier

The operation of both up and downdraught gasifiers is influenced by the morphological, physical and chemical properties of the fuel. Lack of bunker flow, slagging and extreme pressure drop over the gasifier are some common problems encountered in these gasifiers.

Fluidized bed gasifier illustrated schematically in Fig. 7.6 is solution of above all problems. In this type of gasifier air is blown through a bed of solid particles at a sufficient velocity to keep these in a state of suspension. The bed is originally externally heated and the feedstock is introduced as soon as a sufficiently high temperature is reached. The fuel particles are introduced at the bottom of the reactor, very quickly mixed with the bed material and almost instantaneously heated up to the bed temperature. As a result of this treatment the fuel is pyrolysed very fast, resulting in a component mix with a relatively large amount of gaseous materials. Further gasification and tar-conversion reactions occur in the gas phase. Most systems are equipped with an internal cyclone in order to minimize char blow-out as much as possible. Ash particles are also carried over the top of the reactor and have to be removed from the gas stream if the gas is used in engine applications.

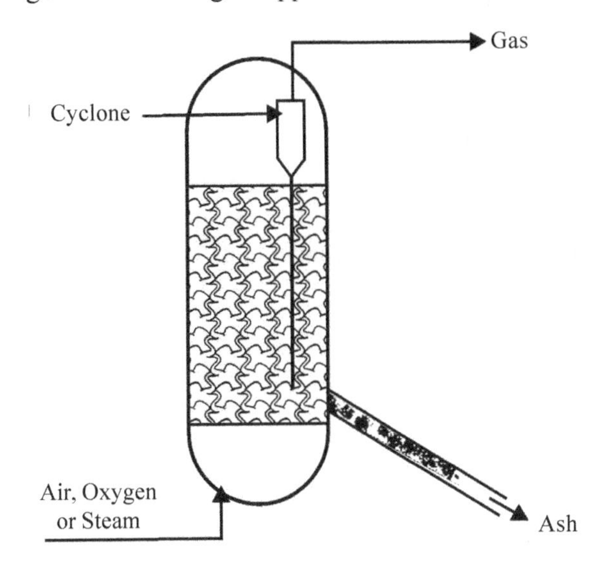

Fig. 7.6: Fluidized Bed Gasifier

7.3.2.1 Bubbling Fluidized Bed Gasifier

Gasification of different biomass feedstock in bubbling fluidized bed reactor takes place over hot bed of inert material such as sand, dolomite etc. under high pressure with fluidizing medium of air, oxygen or steam. These gasifiers are easy in design and in operation as compared to circulating fluidized bed gasifier. These gasifiers are generally designed to operate at low gas velocity of below 1m/s. Solid particles while moving along the gas flow are separated from the gas in the cyclone and collected at the bottom of reactor. Biomass conversion takes place in the bubbling bed region which gives lesser time for tar conversion in the reactor. These gasifiers are capable to operate at higher temperature of 850°C which results in to higher thermal breakdown of biomass and increase in the quantity of syngas production. However carbon conversion efficiency is lower as compared to circulating fluidized bed gasifier[9,12].

7.3.2.2 Circulating Fluidized Bed Gasifier

There are numbers of biomass gasifier installed across the country and most of them are fixed bed type which are not suitable for continuous long run. Though downdraft gasifier found suitable for electricity generation but the producer gas leave from it are having very high temperature. It increases excessive cooling load and considerable heat lost reduce the overall efficiency of the gasifier.

Circulating fluidized bed (CFB) gasifer is advanced and emerging technology as compared to fixed bed type gasifier. Producer gas generated using CFB have low tar, low ash and moderate gas temperature, ultimately reduced gas cooling load and increase overall conversion efficiency.

7.4 Producer Gas and It's Constituents

Producer gas is nothing but it is the mixture of combustible and non-combustible gases. The quantity and quality of producer gas highly depends on the type of feedstock and gasifier operating condition. The constituents of producer gas are listed in Table 7.1

Table 7.1: Constituents of Producer Gas

Carbon Monoxide (CO)	15 - 30%
Hydrogen (H_2)	10 - 20 %
Methane (CH_4)	2 - 4%
Nitrogen (N_2)	45 - 60%
Carbon Dioxide (CO_2)	5 - 15%
Water Vapour (H_2O)	6 - 8%
Calorific Value(Higher Heating Value)	4.5 - 6 MJ/m^3

Carbon monoxide gas is toxic in nature and possesses higher octane number of 106. Higher octane numbers retard the ignition speed. The octane number of Hydrogen in producer gas is varied in the range of 60 - 66 and it improve the ignition speed. Amount of available hydrogen and methane in producer gas are responsible for higher heating value. During the gasification process, air is act as gasification media. Atmospheric air having 78.09% Nitrogen and 20.95 % oxygen, during gasification process Nitrogen act as inert gas that is why the Nitrogen percentage in the producer gas is higher than other gases. If the percentage of carbon dioxide gas in producer gas increases, it seems that the gasification process leading to complete combustion. Higher moisture content reflects the higher moisture content in the feedstock.

7.5 Gasification Process and Reaction Chemistry

The gasification process is nothing but it is a thermochemical conversion process through which soil biomass converted into gaseous fuel. There are four steps taking place in entire process i.e., Drying, Pyrolysis, Oxidation, and Reduction.

7.5.1 Drying

Biomass feed prepared for gasification are normally sun dried and having moisture content about 10-15 %, sometime partially dried biomass may be used which have moisture content upto 25%. Moisture content or water present in the biomass feed removed above 100 °C and feedstock completely bone dry. Temperature in this zone ranging during gasification process is about 100-180 °C. There is no thermal decomposition of biomass taking place in this zone.

7.5.2 Pyrolysis

The biomass is heated in absence of air at elevated temperature about 180 to 700 °C. Biomass thermal decomposed and converted into solid, liquid and gaseous fuel. Mixture of combustible gases are condensable in nature, if condensed properly then liquid fuel i.e., biooil or pyrolytic oil can be produced.

7.5.3 Oxidation

Desired air for partial combustion are introducing in this zone. Temperature varied in the rage of 700-1500 °C. Heterogeneous thermal reaction takes place between air and carbon produced in pyrolysis zone. Most of thermo chemical reactions are exothermic in nature. Carbon react with air produce carbon dioxide and heat on the other hand hydrogen react with air and produce water vapour. The main reactions are-

$$C + O_2 \rightarrow CO_2 + 406 \text{ (MJ/kmol)}$$

$$H_2 + {}^1/_2\, O_2 \rightarrow H_2O + 242 \text{ (MJ/kmol)}$$

7.5.4 Reduction

Numbers of thermo chemical reactions takes place in absence of air. These reactions lower the zone temperature and it sustain at 700 - 1100 °C. The principal reactions are as follows-

$$CO_2 + C \rightarrow 2CO - 172 \text{ (MJ/kmol)}$$

$$C + H_2O \rightarrow CO + H_2 - 131 \text{ (MJ/kmol)}$$

$$CO_2 + H_2 \rightarrow CO + H_2O + 41 \text{ (MJ/kmol)}$$

$$C + 2H_2 \rightarrow CH_4 + 75 \text{ (MJ/kmol)}$$

7.6 Factor Affecting the Biomass Gasification Process

The quality of produce gas highly depends of biomass feedstock. The following parameter to be monitor to get desire quality of producer gas:

7.6.1 *Gasification Media*

Gas composition and its calorific value depend on biomass composition, stability of oxidation zone, temperature at different zone, and gasification media. Role of gasification media on its composition and its calorific value are listed in Table 7.2

Table 7.2: Gasification media, Products and Calorific Value of Producer Gas.

Media	Products	Calorific value of producer gas
Air	CO, CO_2, H_2, CH_4, N_2 and tar	5 – 6 MJ / Nm^3
Oxygen	CO, CO_2, H_2, CH_4, and tar	10 – 12 MJ / Nm^3
Steam	CO, CO_2, H_2, CH_4, and tar	15-20 MJ / Nm^3

7.6.2 *Moisture Content*

Biomass feedstock to be properly dry and kept moisture content below 10-15 % percent. A higher moisture content reduced the temperature oxidation zone. Owing to such issue incomplete cracking of hydrocarbons lowering the calorific value of producer gas.

7.6.3 *Ash Content*

The amount of ash in different types of feedstock varies widely (0.1% for wood and up to 15% for some agricultural products) and influences the design of the ash removal and handling system.

The oxidation temperature is often above the melting point of the biomass ash, leading to clinkering in the grate and subsequent feed blockages. Clinker is a problem for ash contents above 5%, especially if the ash is high in alkali oxides and salts which produce eutectic mixtures with low melting points[13].

7.7 Equivalence Ratio (ER)

Equivalence ratio[13] is one of the most important parameters, which affect the gasification process. Equivalence ratio is the ratio of the actual air to the stoichiometric air/fuel ratio. Its value is 1 for the complete combustion and 0 (zero) for pyrolysis or carbonisation, and for gasification its values typically ranging from 0.2 to 0.4. Several problem associated including excessive char formation if ER < 0.2. Excessive CO_2 and H_2O generation will take place if ER > 0.4

$$ER = \frac{\text{Actual air}}{\text{Stoichiometric air for complete combustion}}$$

7.8 Turndown Ratio

It is the ratio of highest gas generation by a gasifier to the lowest gas generation in actual.The value of trundown ratio of gasifier for vehicle operation is about 3 for Imbert gasifier without insulation while it is about 18 for well insulated gasifier. For running irrigation pumps and electricity generation it is not much important as they running on full load[10].

7.9 Cold and Hot Gas Efficiency

Cold gas efficiency is the ratio of thermal energy of produce gas to the thermal energy of feed stock

$$Cold\ gas\ efficiency = \frac{Heating\ value\ of\ produce\ gas}{Heating\ value\ of\ feedstock} \times 100\ \%$$

Hot gas efficiency: The gas is not cooled before combustion and the sensible heat is also used.

$$Hot\ gas\ efficiency = \frac{Heating\ value\ of\ produce\ gas + Sensible\ heat}{Heating\ value\ of\ feedstock} \times 100\ \%$$

7.10 Capacity of the Gasification System

The gasification system is to be designed on the basis of process heat required kJ per hour

The following assumptions to be made for design:

The hot gas efficiency of the gasification system	=	60%
Burner efficiency	=	60%
Specific gasification rate	=	150 kg $h^{-1}m^{-2}$
Calorific value of gas	=	4600kJ m^{-3}
Calorific value of feed stock	=	12.5 MJ kg^{-1}
Gas output from wood chip (Theoretically)	=	2.2 $m^3\,kg^{-1}$

7.10.1 Feedstock Consumption Rate

The consumption rate of feedstock was calculated using the following assumptions:

Calculation of Feedstock Consumption Rate

$$\text{Feedstock consumption rate} = \frac{\text{Gas output x Calorific value of gas}}{\text{Hot gas efficiency x Calorific value of feedstock}}$$

7.10.2 Dimension of the Reactor Shells

It is calculated by using the following formula:

$$\text{Reactor cross sectional area} = \frac{\text{Feedstock consumption rate}}{\text{Specific gasification rate}}$$

7.10.3 Height of the Reactor

The height of the reactor was decided on the basis of required feedstock holding capacity and the duration of operation of the system.

$$\text{Volume occupied by wood chips} = \frac{\text{Holding Capacity}}{\text{Bulk density of wood chips}}$$

$$\text{Height of wood chips holding column} = \frac{\text{Volume occupied by wood chips}}{\text{Reactor cross sectional area}}$$

The height of the reactor to be fixed in order to accommodate grate, and provide space for ash collection at the bottom.

7.10.4 Critical Thickness of Insulation

Contrary to common belief that addition of insulating material on a surface always brings about a decreasing in the heat transfer rate, there are instances when the addition of insulation to the outside surfaces of walls (geometries which have non-constant cross sectional area) does not reduce the heat loss. In fact, under certain circumstances, it actually increases the heat flow up to a certain thickness of insulation. The insulation thickness at which resistance to heat flow minimum is called critical thickness of insulation. The critical thickness, designed by R is dependent only on thermal quantities k and h_o

$$R = \frac{k}{h_o}$$

where,

R - Critical thickness of insulation, m

k - Thermal Conductivity of material, W/m K

h_o - Coefficient of convective heat transfer, W/m^2 K

7.10.5 Design of the Burner

The producer gas burner is of simple aspirated type. Various details were worked out based on the gas flow rate which are as follows:

$$\text{Velocity of the gas at the inlet of burner} = \frac{\pi \times D \times N}{60} \; m\, s^{-1}$$

Where,

D – Diameter of the blower, m

N–Revolution per minute of blower

$$\text{Area of the cross section of the gas inlet } (A_b) = \frac{\text{Gas flow rate through burner}}{\text{Velocity of gas}}$$

$$\text{Diameter of inlet pipe of the burner}(d_b^{\,2}) = \frac{\text{Gas flow rate through burner} \times 4}{(\text{Velocity of gas} \times 3600 \times \pi)}$$

Diameter of Inlet pipe of the burner =

$$(d_b) = \sqrt{\frac{\text{Gas flow rate through burner} \times 4}{(\text{Velocity of gas} \times 3600 \times \pi)}}$$

The diameter of the burner can be designed according to the size of the vessels.

7.11 Gasifier and Its Applications

7.11.1 Power Applications

Gas engine can be run on100 percent producer gas. Biomass required to produce of 1 kWh of power from gas engine is varied from 1.12 to 1.5 kg and it depends on the type of ignition. On the other hand engine running on dual fuel mode (diesel + producer gas) approximately 70 to 80 percent diesel can be replaced using producer gas. In dual fuel mode the engine likely to de-rated by 20 to 30 percent on rated capacity basis. Biomass and diesel fuel required to produce 1 kWh of power are 0.9-1.2 kg and less than 0.1 liter respectively.

7.11.2 Thermal Applications

Biomass gasifier found most suitable for thermal applications. There is no requirement cleaning and cooling of producer gas, it can directly be combust using produce gas burner. It is a viable option to replace the convention fossil fuel such as diesel, light diesel oil, furnace oil and LPG. The biomass required to replace one liter of diesel and light diesel oil are in the range of 3.5 to 4.25 kg respectively, where as one kilogram of LPG and furnace oil are in range of 4 to 4.5 kg and 4 to 5 kg respectively.

7.12 Gasifier Operation

7.12.1 Starting of Gasifier

The open core downdraft biomass gasifier was design and developed by Sardar Patel Renewable Energy Research Institute and supply to Department of Renewable Energy Engineering, MPUAT, Udaipur to install at Agro based industries under AICRP of EAAI. Before starting the gasifier system leak proof test was conducted (Fig.7.7). Thereafter gasifier is loaded with 25 kg of charcoal over the grate, with 10 kg of wood saving and 40 kg of sized wood. At initial stage, the starting blower is switched on and ignited torch is kept near the air entry ports of the main reactor so that the oxidation zone is established (Fig.7.8). Then the gas at the flare burner is examined for combustion using an ignited torch. After flaring the gas (Fig. 7.9), the main blower is switched on and the gas is allowed to burn in the furnace. Necessary instruments were attached to this system to measure temperature, pressure difference, and orifice meter for gas flow rate. A Junkar gas calorimeter was also used to calculate calorific value of producer gas.

Fig.7.7: Leakage Test of Gasifier

Fig. 7.8 a&b : Igniting the Gasifier

Fig.7.9: Gasifier with Flame

7.12.2 Findings

The babul wood (*(Prosopis juliflora)* was used for running the gasifier. The physical and thermal properties of the biomass are given in Table 7.3.The initial test of gasifier was carried out as per SPRERI guidelines, its results are given in Table 7.4. The calorific value of producer gas was also calculated with the help of Junkar gas calorimeter, the results are given in Table 7.5. The test run of gasifier flame temperature, pressure drop across the gasifier, at orifice plate and flow rate are given in Table 7.6.

Table 7.3: Physical and Thermal Properties of Biomass

S.No.	Characteristic		Biomass (babul wood) *(Prosopis juliflora)*
1.	Size (mm)	:	25-40
2.	Length (mm)	:	35-85
3.	Bulk density (kg m^{-3})	:	395
4.	Angle of slide (deg)	:	15
5.	Moisture content (% wb)	:	10.2
6.	Volatile matter (% db)	:	83.42
7.	Ash content (%db)	:	1.05
8.	Fix carbon (%db)	:	15.53
9.	Calorific value (kcal kg^{-1})	:	3895

Table 7.4: Initial Test Run (ITR) of Open Core Downdraft Gasifier

Fuel	:	Vilayti babul (*Prospour juliflroa*)
Moisture content	:	10.2% wb
Calorific value	:	3895 kcal/kg (wet basis)
Load before / during test		
Charcoal added	:	25 kg
Wood saving	:	10 kg
Fuel added	:	40 kg
Flaring		**Time**
Nozzle ignition	:	8:25 am
Gas flaring	:	8.30 am
Ambient conditions	:	24 °C

Table 7.5: Junkar Gas Calorimeter

S.No.	Gas Temperature at Junkar burner inlet	Water temp. at inlet °C	Water temp. at outlet °C	Water column liter	Gas column liter	Gas calorific value at ambient kcal /m^3
1	26	26	32	0.92	5.6	986

Example 7.1: A downdraft biomass gasifier installed at boys hostel to run the canteen. Canteen owner is using groundnut shell briquettes as fuel to run gasifier. The ultimate analysis of groundnut shell is as follows:

C = 41.10 %

H = 4.8 %

N = 1.6 %

O = 39.2 %

S = 0.04%

Table 7.6: Result of Run

S.No.	Time	Activity	$T_{raw\ gas}$ °C	T_{flame} °C	Gas temp burner°C	ΔP gasifier cm WG	ΔP orifice cm WC	Biomass added kg	Gas flow rate M³/hr
1.	10:30 am	Gas burned inside the oven	210	—	85	8	15.0		60
2.	11:10 am		215	748	87		18.0	—	65
3.	11:40 am	Top up wood	245	—	98	12	26.0	45 kg	78
4.	12:10am	Grate agitate	285	—	115	15	32.0	—	90
5.	12:40am		305	—	118	9	40.2	35 kg	95
6.	01:10 pm		334	—	125			—	
7.	01:40 am	Grate agitate	370	—	122	18	45.0	—	110
8.	02:10 am	Top up wood	383	750	125	7		60 kg	
9.	02:40 am	Grate agitate	381	—	120	16	44.5	—	
10.	03:10 am		393	—	120	9	45.0	—	110
11.	03:40 pm		391	752	121	13	44.21	—	
12.	04:10 pm	Grate agitate	395	—	122	16	45	—	
13.	04:40 pm	Top up wood	392	—	123	8	44.1	60 kg	
14.	05:10 pm		394	750	125		45	—	
15.	05:40 pm	Grate agitate	391	—	—	15	—	—	
16.	06:10 pm		393	752		9	44.9		
17.	06:40 pm		392			11	45	—	110

Determine the chemical formula for groundnut shell and desired air fuel ratio for complete combustion. Further, estimate air required for gasification if only 35% of the stoichiometric air used during gasification. Assuming air density is about 1.2 kg/m³.

Solution

Chemical formula

$$C = \frac{41.10}{12} = 3.42$$

$$H = \frac{4.8}{1} = 4.8$$

$$N = \frac{1.4}{14} = 0.10$$

$$O = \frac{39.2}{16} = 2.45$$

$$S = \frac{0.04}{32} = 0.0012$$

Thus chemical formula of groundnut shell is as follows:

$C_{3.42}H_{4.8}O_{2.45}N_{0.10}S_{0.0012}$

Air required for complete combustion of groundnut shell briquettes:

$$C_{3.42}H_{4.8}O_{2.45}N_{0.10}S_{0.0012} + 3.39\ [(O_2 + 3.76N_2)]$$

$$\rightarrow 3.42CO_2 + 2.4H_2O + 3.39 \times 3.76N_2 + 0.0012S$$

$$\frac{Air}{Fuel} = \frac{(3.39\times32)+(3.39\times3.76\times28)}{(12\times3.42)+(1\times4.8)+(16\times2.45)+(14\times0.1)+(32\times0.0012} = 5.38$$

For complete combustion of biomass 5.38 kg of air per kg of fuel is required. For gasification only 35% of desired air is used. Therefore actual air required for gasification is:

$5.38 \times 0.35 = 1.883$ kg

Air fuel ratio for gasification:

$\frac{1.883}{1.2} = 1.56\,m^3$ air per kg of groundnhut shell brequitte

Example 7.2: Produce gas combusted in boiler to generate steam with 80 % thermal efficiency. Steam expanded in steam turbine from 500 °C to 100 °C to produce electricity. Calculate the overall efficiency of system.

Solution

Given

Thermal efficiency: 80%

Steam temperature at turbine inlet: 500 °C = 773 K

Steam temperature at turbine outlet: 100 °C = 373 K

Ideal turbine efficiency:

$$\eta = \left(1 - \frac{373}{773}\right) \times 100 = 51.76\ \%$$

Therefore, overall system efficiency:

ηoverall = 0.80 × 0.5176 = 41.39 %

Example 7.3: Electricity consumption in a food processing industry is about 5 MW. Owner of industry installed a downdraft biomass gasifier to meet electricity demand. Gasifier based power plant operated 300 days in a year and 24 hour per day. Calorific value of biomass used is 16 MJ/kg. The conversion efficiency is about 18%. Calculate the area required for biomass production for smooth running of power plant, if biomass yield is about 30 tons per hectare.

Solution

Energy require : 5 MW

Conversion efficiency : 18%

Biomass yield : 30 tons per hectare

Operation 300 days, 24 hours per day

$$\text{Energy input} = \frac{\text{Energy Required}}{\text{Conversion efficiency}} = \frac{5}{0.18} = 27.77\ \text{MW}$$

Area requires for cultivating biomass for uninterrupted power supply:

$$27.77\frac{\text{MJ}}{\text{s}} \times 300\ \text{days} \times 24\ \text{hours} \times 3600\ \text{second} \times \frac{1\ \text{kg}}{16\ \text{MJ}} \times \frac{1}{30000}\frac{\text{hectare}}{\text{kg}} = 1499.58\ \text{hectare}$$

Example 7.4: Calculate the biomass requirement per hour to replace 50 kg of Light Diesel Oil (LDO) operated steam generated. Assume calorific value of LDO and biomass are 40 MJ/kg and 19 MJ/kg respectively, and overall efficiency for both gasification and LDO operated steam generated system is about 20 %.

Solution

Energy supplied by LDO system

$$= \frac{50\,kg \times 40\dfrac{MJ}{kg}}{0.20} = 10,000\frac{MJ}{hour}$$

Energy input desired from biomass:

$$= \frac{10,000\,MJ}{0.20} = 50,000\frac{MJ}{hour}$$

Biomass required supplying desired energy:

$$= \frac{50,000\,MJ}{19\dfrac{MJ}{kg}(calorific\,value\,of\,biomass)} = 2631\frac{kg}{hour}$$

Example 7.5: Calculate the diameter and height of a downdraft biomass gasifier to supply 500 kW thermal energy to generate process heat for 4 hours. Biomass gasifying at 60% efficiency with specification gasification rate of 130 kg h^{-1} m^{-2}. Assume heating value of biomass and producer gas are 16.5 MJ kg^{-1} and 5.2 MJ Nm^{-1} respectively.

Solution

Themal output = Producer gas flow rate × Heating value of producer gas

$$\text{Producer gas flow rate} = \frac{\text{Thermal output}}{\text{Heating value of producer gas}}$$

$$\text{Producer gas flow rate} = \frac{500}{1000}\frac{\text{MJ}}{\text{s}} \times \frac{1}{5.2}\frac{\text{Nm}^{-3}}{\text{MJ}}$$

$$\text{Producer gas flow rate} = 0.096\frac{\text{Nm}^{-3}}{\text{s}}$$

Biomass required to generate desired producer gas flow rate

$$\text{Fuel consumption rate} = \frac{\text{Thermal power output}}{\text{Heating value of biomass} \times \text{Heating value of producer gas}}$$

$$\text{Fuel consumption rate} = \frac{500}{1000}\frac{\text{MJ}}{\text{s}} \times \frac{1}{16.5}\frac{\text{kg}}{\text{MJ}} \times \frac{1}{0.6} = 0.050\frac{\text{kg}}{\text{s}}$$

Fuel consumption rate $= 181.81\frac{kg}{h}$

Internal diameter of gasifier reactor

$$\text{Cross sectional area of reactor} = \frac{\text{Fuel Consumption rate}}{\text{Speific gasification rate}}$$

$$\text{Cross sectional are of reactor} = 181.81\frac{kg}{h} \times \frac{1}{130}\frac{h\,m^2}{kg} = 1.39\ m^2$$

$$\text{Diamenter of reactor} = \sqrt{\left(\frac{1.39 \times 4}{\pi}\right)} = 1.315\ m$$

If gasifier operated in single feeding, 727.24 kg biomass to be loaded. Assuming bulk density of biomass is 350 kg m^{-3}

$$\text{Height of wood column} = \frac{\text{Fuel consumption rate} \times \text{Duty hour}}{\text{Bulk Density of Biomass}}$$

$$Height\ of\ wood\ column = \frac{181.81 \times 4}{350} = 2.07\,m$$

Example 7.6: Calculate the Stoichiometric air required for the solid fuel having the chemical composition: $C_{3.33}H_6N_{0.071}O_{2.38}$

Solution: The molecular weight of the C =12, H=1, N=14 and O=16

So, Actual Proportion of the C, H, N, O in fuel are:

$C = 3.33 \times 12 = 39.96\ \% = 0.3996$

$H = 6 \times 1 = 6\ \% = 0.06$

$N = 0.071 \times 14 = 0.994\ \% = 0.00994$

$O = 2.38 \times 16 = 38.08\% = 0.3808$

Out of all of the above components the oxidation reaction will takes place with Carbon and Hydrogen Only

But, In the atmosphere $N_2 = 79\%$ and $O_2 = 21\ \%$

$$\text{Therefore one Oxygen molecule} = O_2 + \frac{79}{21}N_2 = O_2 + 3.76N_2$$

Now For the Carbon,

$C + (O_2 + 3.76N_2) \rightarrow CO_2 + 3.76N_2$

adding the moecular weight

$$12 + (2 \times 16 + 3.76 \times (2 \times 14)) \rightarrow 44 + 3.76 \times 2 \times 14$$

$$12C + 320_2 + 105.28N_2 \rightarrow 44CO_2 + 105.28N_2)$$

Now adding the proportionate in to the equation

$$0.3996C + 0.3996 \times \frac{32}{12} O_2 + 0.3996 \times \frac{105.28}{12} N_2$$

$$\rightarrow 0.3996 \times \frac{44}{12} CO_2 + 0.3996 \times \frac{10.5.28}{12} N_2$$

$$0.3996\ C + 1.065\ O_2 + 3.50N_2 \rightarrow 1.465\ CO_2 + 3.50N_2$$

For the Hydrogen,

$$H_2 + \frac{1}{2}(O_2 + 3.76N_2) \rightarrow H_2O + \frac{3.76}{2} N_2$$

$$2 + \left(\frac{1}{2} \times 32\right) + \left(\frac{1}{2} \times 3.76 \times 28\right) \rightarrow 18 + \left(\frac{1}{2} \times 3.76 \times 28\right)$$

$$2H_2 + 16\ O_2 + 52.64\ N_2 \rightarrow 18\ H_2O + 52.64\ N_2$$

$$0.06 + \left(0.06 \times \frac{16}{2}\right) + \left(0.06 \times \frac{52.64}{2}\right) \rightarrow \left(0.06 \times \frac{18}{2}\right) + \left(0.06 \times \frac{52.64}{2}\right)$$

$$0.06\ H_2 + 0.48\ O_2 + 1.58\ N_2 \rightarrow 0.54\ H_2O + 1.58\ N_2$$

Now, adding all the composition on Left Hand Side and,

$$0.3996\ C + 0.06\ H_2 + (0.00994\ N_2 + 0.3808\ O_2) + (1.065\ O_2 + 3.50\ N_2 + 0.48\ O_2 + 1.58N_2)$$

$$\rightarrow 1.465\ CO_2 + 0.54\ H_2O + 3.50\ N_2 + 1.58\ N_2$$

$$+ (0.00994\ N_2 + 0.3808\ O_2)$$

$$0.3996\ C + 0.06\ H_2 + (0.00994\ N_2 + 0.3808\ O_2) + 1.545\ O_2 + 5.08\ N_2$$

$$\rightarrow 1.465\ CO_2 + 0.54\ H_2O + 3.50\ N_2 + 1.58\ N_2$$

$$+ (0.00994\ N_2 + 0.3808\ O_2)$$

Total amount of air required ($O_2 + 3.76\ N_2$) = 1.545 + 5.08 = 6.625

For 1 kg of fuel we required 6.625 kg of total air

References

[1] Verma R, Singh M. Symmetrical analysis of biomass gasification techniques. MR Int J Eng Technol 2013;5(1):17–24.

[2] Hanne R, Kristina K, Alexander K, Arunas B, Pekka S, Matti R, Outi K, Marita N. Thermal plasma-sprayed nickel catalysts in the clean-up of biomass gasification gas. Fuel, 2011; 90: 1076-1089.

[3] McKendry P.Energy production from biomass (part 2): conversion technologies. Bioresource Technology 2002; 83: 47-54.

[4] Özbay N, Pütün AE, Pütün E. Structural analysis of bio-oils from pyrolysis and steam pyrolysis of cottonseed cake. Journal of Analytical and Applied Pyrolysis. 2001 Jun 1;60(1):89-101.

[5] Balat M, Balat M, Kýrtay E, Balat H. Main routes for the thermo-conversion of biomass into fuels and chemicals. Part 2: Gasification systems. Energy Conversion and Management 2009; 50: 3158–3168.

[6] Damartzis T, Zabaniotou A. Thermochemical conversion of biomass to second generation biofuels through integrated process design—A review. Renewable and Sustainable Energy Reviews 2011;15: 366–378.

[7] Rajvanshi AK. Biomass gasification. Boca Raton, Florida, United States: CRC Press; 1986. p. 83–102

[8] Biomass Gasification Technology and Utilisation http://cturare.tripod.com/his.htm (retrieve on December 12.2019.

[9] Singh J. Management of the agriculture biomass on decentralized basis for producing sustainable power in India. Journal of Cleaner Production 2016: 1-16.

10. Reed TB, Das A. *Handbook of biomass downdraft gasifier engine systems*. Biomass Energy Foundation 1988.

[11] Srivastava T. Renewable energy (gasification). Adv Electron Electr Eng 2013;3(9):1243–50.

[12] Latif A. A study of the design of fluidized bed reactors for biomass gasification[Ph.D. thesis]. London: University of London; 1999.

[13] Kýrtay E. Recent advances in production of hydrogen from biomass. *Energy Conversion and Management 2011;52*(4): 1778-1789.

References

[1] [illegible] analysis of biomass gasification techniques. [illegible] 17–24.

[2] [illegible] Alexander K, [illegible] two-staged [illegible] in the clean-up of biomass gasification gas. Fuel, 2011, [illegible]

[3] [illegible] Energy produced from biomass [illegible] gasification technology [illegible]

[4] [illegible] thermal analysis [illegible] Analytical and Applied Pyrolysis, [illegible]

[5] [illegible] conversion of biomass into [illegible]

[illegible]

8

Biodiesel Production

8.1 Introduction

Rising petroleum prices, increasing threat to the environment from exhaust emissions and global warming have generated an intense international interest in developing alternative non-petroleum fuels for engines. The use of vegetable oil in internal combustion engines is not a recent innovation. Rudolf Diesel (1858-1913), creator of the diesel cycle engines, used peanut vegetable oil to demonstrate its invention in Paris in 1900. In 1912, Diesel said, *"The use of vegetable oils as engine fuel may seem negligible today. Nevertheless, such oils may become, in the passing years, as important as oil and coal tar presently."* Nowadays, it is known that oil is a finite resource and that its price tends to increase exponentially, as its reserves have being decreasing[1].

India ranks sixth in the world in total energy consumption and needs to accelerate the development of the sector to meet its growth aspirations. The country, though rich in coal and abundantly endowed with renewable energy in the form of solar, wind, hydro and bio-energy has very small hydrocarbon reserves (0.4 % of the world's reserve). India, like many other developing countries, is a net importer of energy, more than 25 percent of primary energy needs being met through imports mainly in the form of crude oil and natural gas. The rising oil import bill has been the focus of serious concerns due to the pressure it has placed on scarce foreign exchange resources and is also largely responsible for energy supply shortages. The sub-optimal consumption of commercial energy adversely affects the productive sectors, which in turn hampers economic growth[2].

8.2 Biodiesel Chemistry

Biodiesel is a mixture of methyl esters of long chain fatty acids like lauric, palmitic, stearic, oleic, and so on. It is produced by the transesterification of animal fats and vegetable oils – all of which belong to a group of organic esters called triglycerides. Typical examples are rape/canola oil, soyabean oil, sunflower oil, palm oil and its derivatives, etc. from vegetable sources, beef and sheep tallow and poultry oil from animal sources and also from used cooking oil. The chemistry is basically the same irrespective of the feedstock. The chemistry of biodiesel is given in Fig. 8.1

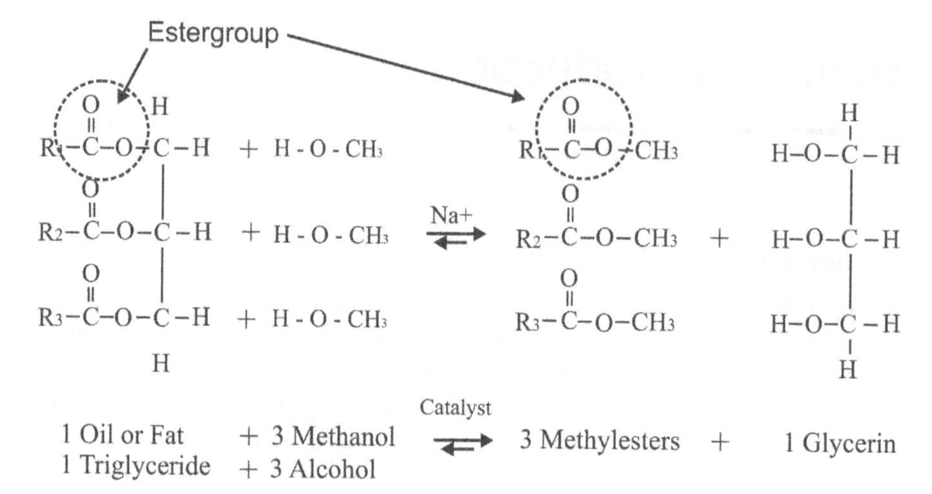

Fig. 8.1: Biodiesel Chemistry

The abbreviation R_1, R_2 and R_3 are symbolic representations of the fatty acid chains, which can vary in molecular chain length from typically C8 to C22, and also their degree of unsaturation. An example is **Oleic acid**, which has 18 carbon atoms and one double bond:

CH_3-CH_2-CH_2-CH_2-CH_2-CH_2-CH_2-CH_2-CH=CH-CH_2-CH_2-CH_2-CH_2-CH_2-CH_2-CH_2-COOH

or more simply: CH_3-$(CH_2)_7$CH=CH-$(CH_2)_7$COOH

The abbreviation C(18:1) is also often used.

8.3 Biodiesel Process

The production of biodiesel involves intensively mixing of methanol with oil or fat in the presence of a suitable catalyst, and then allowing the lighter methyl ester phase to separate by gravity from the heavier glycerol phase. However, as with most organic reactions the degree of conversion depends on the equilibrium reached as well as the influence of other reactions. Achieving product quality is also very important. An example of process configuration designed to optimize the production of biodiesel is illustrated in Fig.8.2

Process flow chart: Implementing the previously described processes results in the process flow shown in Fig. 8.2

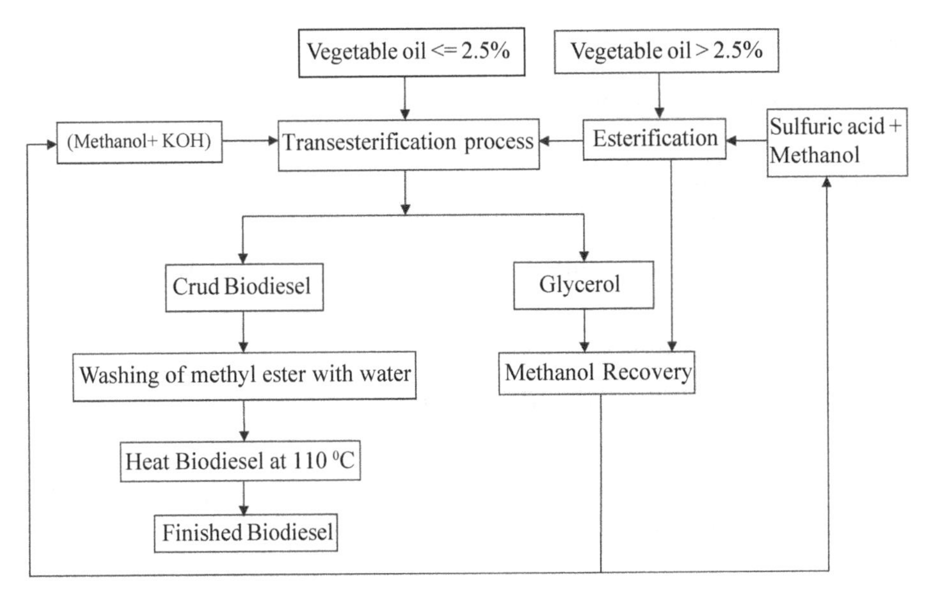

Fig. 8.2: Process Flow for Biodiesel Process

The key features of Biodiesel Process are:

- Technology applicable to multiple feedstocks
- Continuous process at atmospheric pressure and nominally 60°C temperature
- Dual Reactor System operating with a patented Glycerin Cross Flow configuration for maximized conversion
- Reaction using excess methanol, but with full methanol recycle to avoid any losses
- Closed loop water wash recycle to minimize waste water generation
- Clear phase separation by special gravity process (no centrifuges necessary)
- Based on suitable feedstock, this produces biodiesel to current world standards
- Raw Glycerin to BS 2621.

The crude glycerin can be further upgraded to pharmaceutical glycerin standard EU Pharmacopoeia 99.5 by distillation, bleaching if required, and vacuum drying.

8.4 Key Issues in the Manufacturing of Biodiesel

8.4.1 Quality of Feedstock

The performance of the process and the quality of the Biodiesel produced during transesterification are dictated by the quality of feedstock. Therefore, it is important to feed an oil to the process that has been pretreated:

- To remove water as this causes dissociation of the catalyst
- To remove free fatty acids as they will produce soaps which interfere with phase separation and distillation
- To remove insoluble material, sediments and carbon residues as they can fail the corresponding biodiesel specification tests
- To remove phosphatides as they interfere with phase separation and they generate slime and gumming characteristics

8.4.2 Nature of Feedstock

Nature of feedstock is also one of the important characteristics for production of biodiesel. Not all vegetable oils can be converted to biodiesel. Certain methyl esters, particularly those from saturated fatty acids, display high melting temperatures, and even when blended with many other methyl esters, this characteristic may lead to failing certain biodiesel specification tests. All diesel fuel is filtered before being injected into a diesel engine. If the biodiesel or the biodiesel blend is sufficiently cold then some of these particular methyl esters will crystallize out of solution and cause the filter to plug up, thus starving the diesel engine. The particular test is termed as the "Cold Filter Plug Point" and the European Standard EN 14214 has different Classes depending on climate and the time of year - Class A for warm climates the CCPP is 5°C (max) but for very cold climates Class F has a CFPP of –20°C (max).

Similar Tests such as "Cloud Point" and "Pour Point" also provide a means of differentiating between different biodiesels. For example the Pour Point of Rape Methyl Ester (RME) is –15C, Soy Methyl Ester (SME) is –3°C and Tallow Methyl Ester (TME) +16°C. Obviously TME will fail the test first, but there are several ways of overcoming this problem, such as limiting the amount of biodiesel added to say 2% (B2), adding a Pour Point Suppressant or fractionating the biodiesel product to reduce the level of the responsible methyl esters.

8.4.3 Influence of Feedstock on Glycerin Quality

Glycerin is used extensively throughout the pharmaceutical and cosmetic industries. Particular emphasis is placed on sourcing the glycerin from non-animal feedstocks, such that it can be termed Kosher quality. Such glycerin commands significantly higher prices than that sourced from tallow or other animal sources. It is even becoming more important to avoid potentially contaminating vegetable based glycerin with any other material. With the coming growth in biodiesel production pressure will be put on glycerin prices and finding markets for non-kosher glycerin. The sale of glycerin is an important revenue stream for the economics of biodiesel production.

8.5 Characteristics of Biodiesel

Biodiesel as automotive fuel has similar properties to petrodiesel and as such can be directly used in existing diesel engines with no or minor modifications. It can be used alone or mixed in any ratio with petrodiesel. The most common blend is B20, a mix of 20% biodiesel with 80% petroleum diesel. Biodiesel has 11% oxygen by weight and essentially contains no sulphur or aromatics.

8.5.1 Physical Properties

Some of the physical characteristics of biodiesel are given in Table-8.1.

Table 8.1: Physical Properties of Biodiesel

S.No.	Properties	Values
1.	Specific gravity	0.88
2.	Viscosity @ 20 (centistokes)	7.5
3.	Cetane Index	49
4.	Cold filter Plugging Point (°C)	-12
5.	Net Heating Value (Kilojoules/Liter)	33,300

8.5.2 Biodiesel Specifications

The key components, which determine the quality of biodiesel are monoalyl esters, dialkyl esters, residual vegetable oil, free glycerin, reactant alcohol, free fatty acids and the residual catalyst. In December 2001, American Society of Testing & Materials (ASTM) issued a specification (D6751) for biodiesel (B100) which is presented in Table-8.2. A provisional specification for B20 biodiesel is also notified by the ASTM known as ASTM PS121 (Table-8.3). Table-8.4 shows a comparison of selected properties of biodiesel and petrodiesel.

Table 8.2: ASTM Specification (D6751) for B100

Property	ASTM Method	Limits	Units
Flash Point	D93	130 min.	Degrees C
Water & Sediment	D2709	0.050 max.	% Volume
Kinematic Viscosity (40 C)	D445	1.9-6.0	mm^2/sec
Sulfated Ash	D874	0.020 max.	% mass
Sulphur	D5453	0.05 max.	% mass
CopperStrip Corrosion	D130	No.3 max.	
Cetane	D613	47 min.	
Cloud Point	D2500	Report	Degrees C
Carbon Residue (100% Sample)	D4530*	0.050 max.	% mass
Acid Number	D664	0.80 max.	Mg KOH/gm
Free Glycerin	D6584	0.020 max.	% mass
Total Glycerin	D6584	0.240 max.	% mass
Phosphorous Content	D4951	0.001 max.	% mass
Distillation Temperature (90% Recovered)	D1160	360 max.	°C

*The carbon residue shall be run out on the 100% sample.

Table 8.3: ASTM PS121 for B20

Property	ASTM Method	Limits	Units
Flash Point	D93	100 min.	°C
Water & Sediment	D2709	0.050 max.	% Volume
Kinematic Viscosity (40 C)	D445	1.9-6.0	mm^2/sec
Sulfated Ash	D874	0.020 max.	% mass
Sulphur	D5453	0.0015 max.	% mass
Copper Strip Corrosion	D130	No. 3 max.	
Cetane Number	D613	46 min.	
Cloud Point	D2500	Report	°C
Carbon Residue (100% Sample)	D4530	0.050 max.	% mass
Carbon Residue (Ramsbottom)	D524	0.090 max.	% mass
Acid Number	D664	0.80 max.	Mg KOH/gm
Free Glycerin	D6584	0.020 max.	% mass
Total Glycerin	D6584	0.240 max.	% mass

Table 8.4: Selected Fuel Properties for Petrodiesel & Biodiesel

Fuel Property	Petrodiesel	Biodiesel
Fuel Standard	ASTM D975	ASTM PS121
Fuel Composition	C10-C21 HC	C12-C22 FAME
Lower Heating Value, Btu/gal	131,295	117,093
Kin. Viscosity, @40 C	1.3-4.1	1.9-6.0
Specific Gravity, kg/l @ 60 F	0.85	0.88
Density, lb/gal @ 15 C	7.079	7.328
Water, ppm by wt.	161	0.05% max.
Carbon, wt%	87	77

(Contd.)

Hydrogen, wt%	13	12
Oxygen by dif. Wt%	0	11
Sulphur, wt%	0.05 max.	0.00-0.0024
Boiling Point, °C	188-343	182-338
Flash Point, °C	60-80	100-170
Cloud Point, °C	-15 to 5	-3 to 12
Pour Point, °C	-35 to -15	-15 to 10
Cetane Number	40-55	48-65
Stoichometric Air/Fuel Ratio	15	13.8
BOCLE Scuff, gm	3,600	>7,000
HFRR, microns	685	314

8.5.3 Toxicity of Biodiesel

Impacts of biodiesel on human health is also a significant criteria for including suitability of the fuel for commercial applications. Health effects can be measured in terms of fuel toxicity to the human body as well as health impacts due to exhaust emissions. Tests conducted by the Wil Research Laboratories investigated the acute oral toxicity of pure biodiesel fuel as well as B20 in a single dose study on rats, which concluded that biodiesel is not a toxic and there is no hazards anticipated from ingestion incidental to industrial exposure. The acute oral LD50 (lethal dose) is greater than 17.4-g/kg-body weight, which by comparison is far safer then even table salt. According to National Institute for Occupational Safety & Human Health, a 96-hr. lethal concentration of biodiesel for bluegills was greater than 1000 mg/l and this aquatic toxicity is deemed as insignificant. Other related effects of biodiesel are given below:

- Very mild human skin irritation. It is less than the irritation produced by 4% soap and water solution.
- It is biodegradable. It degrades at least 4 to 5 times faster then conventional diesel fuel.
- Biodiesel has a flash point of about 300 °F well above conventional diesel fuel.
- Spills of biodiesel can decolorize any painted surface if left for long.
- There is no tendency for the mutagenicity of exhaust gas to increase for a vehicle running on biodiesel (20%RSME+80% diesel).

8.6 Prospective Feedstock in India

Oil can be extracted from a variety of plants and oilseeds. Under Indian condition only such plant sources can be considered for biodiesel production which are not edible oil in appreciable quantity and which can be grown on large-scale on wastelands. Moreover, some plants and seeds in India have tremendous medicinal value, considering these plants for biodiesel production may not be a viable and wise option. Considering all the above options, probable biodiesel yielding trees in India are:

(1) *Jatropha curcas* or Ratanjot

(2) *Pongamia pinnata* or Karanj

(3) *Calophyllum inophyllum* or Nagchampa

(4) *Hevea brasiliensis* or Rubber seeds

(5) *Calotropis gigantia* or Ark

(6) *Euphorbia tirucalli* or Sher and

(7) Boswellia ovalifololata.

Among all the above prospective plant for as biodiesel production, Jatropha curcas stands at the top. One hectare Jatropha plantation with 4400 plants under rain fed conditions can yield about 1500 literes of oil. It is estimated that about 3 million hectares plantation is required to produce oil for 10% replacement of petrodiesel. The residue oil cake after extraction of oil from Jatropha can be used as organic fertilizers. It is also estimated that one acre of Jatropha plantation could produce oil sufficient to meet the energy requirement of a family of 5 members and the oil cake left out when used as fertilizer could cater to one acre.

8.7 Economic Feasibility

The expected cost of extracted oil in Indian condition would be Rs. 35-45 per liter. However, the price of commercially available bio-diesel is presently Rs. 45 – 55 per liter. With recent increase in oil prices, it is essential to look for substitutes of fossil fuels for both economic and environmental benefits to the country. The bio-fuel extraction will disseminate technology and provide better energy services at the village level.

8.8 Environmental Benefits

There are number of environmental benefit by the use of biodiesel, which are given as below:

- The use of bio-fuel avoids fossil fuel use and hence avoids CO_2/CO emission in atmosphere
- It is a promising alternative fuels source for future
- Substantial reduction of unburned hydrocarbons, carbon monoxide and particulate matter

- Decrease the solid carbon fraction of particulate matter
- Increase in the green cover as result of plantations would check soil erosion and retain moisture and soil nutrients.
- Positive ecological benefits in terms of lending support to biodiversity, especially since degraded lands are involved

Biofuels also help the environment because the plants grown to make these fuels take greenhouse gases such as carbon dioxide out of the air and fix it in their roots, stems and leaves. Much of this carbon dioxide gets sequestered in the soil, reducing the overall level of carbon dioxide in the atmosphere.

8.9 Biodiesel Production from Caster Seed Oil

India is the world leader in castor seed and oil production and dominates the international castor oil trade. The Indian variety of castor has 48 % oil content of which 42% can be extracted, while the cake retains the rest. India's castor production fluctuates between 0.6 to 1 million tonnes a year. The state of Gujarat accounts for over 80% of India's castor seed production, followed by Andhra Pradesh and Rajasthan. In Gujarat, castor is mainly grown in the Mehsana, Banaskantha and Saurashtra/Kutch. In Andhra Pradesh, the main castor growing areas are Nalgonda and Mahboobnagar districts. Castor is a Kharif crop. Its sowing season of castor is from July to October and the harvesting season is from October to April[3]. India's annually exports around quarter million tons of commercial castor oil, 1,50,000 tons of castor seed extractions (castor meal), and about 20,000 tons of castor seed[4]. The steps involve in biodiesel production from castor seed oil are as follow:

8.9.1 Heating of Oil

In order to speed up the reaction, the castor oil must be heated. The ideal temperature range is 48 °C to 60 °C. The reaction can take days at room temperature and will be inhibited above 60 °C. Heating with electric elements is usually the easiest way to bring the castor oil up to temperature. One litre measured quantity of fresh, moisture free castor oil was poured in the reactor vessel. Castor oil was heated up to 60 °C temperature and continuously stirred at constant slow speed. It is important to stir the castor oil as it is heated. This will result in a more even heating and reduce the temperature of castor oil exposed directly to the heating element.

8.9.2 Mixing of Methanol and Catalyst

The reactants for transesterification process will be used in the following way, proportion in metric units.

Castor (Ricinus communis) oil	1 kg
Anhydrous Methanol	0.300 kg
Potassium Hydroxide	0.020 kg

Another component will be calculated from the formulae,

$M_tOH = 0.300 \times RO$

$KOH = 0.020 \times RO$

Where,

M_tOH – Amount of methanol required, kg

RO - desired amount of Castor oil to be processed, kg

KOH – Amount of KOH required, kg

The purpose of mixing methanol and the catalyst (KOH) is to react the two substances to form Methoxide. The amount of Methanol used should be 30% of the volume of the oil. Methanol and KOH are dangerous chemicals by themselves, with Methoxide even more so. None of these substances should ever touch skin. KOH does not readily dissolve into Methanol. It is best to turn on the mixer to begin agitating the Methanol and slowly pour the KOH in. When particles of KOH cannot be seen, the Methoxide is ready to be added to the oil. This can usually be achieved in 20 –30 minutes.

8.9.3 Draining of Glycerin

After the transesterification reaction, one must wait for the glycerin to settle to the bottom of the container. This happens because Glycerin is heavier then biodiesel. The settling will begin immediately, but the mixture should be left a minimum of eight hours (preferably 12 hours) for settling and separation of glycerin at bottom to make sure all of the Glycerin has settled out. The Glycerin volume should be approximately 20% of the original oil volume. It shows the difference in viscosity and color between the two liquids. The object is to remove only the Glycerin and stop when the biodiesel is reached. Glycerin looks very dark compared to the yellow biodiesel. The viscosity difference is large enough between the two liquids that the difference in flow from the drain can be seen.

8.9.4 Washing of Fuel

The washing of raw/crude biodiesel fuel is one of the most discussed subjects. The purpose is to wash out the remnants of the catalyst and other impurities. There are three main methods:

1. Water wash only (a misting of water over the fuel, draining water off the bottom)
2. Air bubble wash (slow bubbling of air through the fuel)
3. Air/water bubble wash (with water in the bottom of the tank, bubbling air through water and then the fuel)

8.9.5 Air/water Bubble Wash

The upper biodiesel was put into other transparent vessel for washing with equal amount of water and can be drained throughout the washing process. Bubble washing with water will remove most of the impurities from the biodiesel. Water by 100 per cent volume ratio was used for washing. Water was gently added to the biodiesel so the two liquids do not mix. The water will settle to the bottom of the biodiesel in washing tank. Turn on the aquarium air pump and let it bubble for 1-2 hours. The biodiesel should be washed 2 or 3 more times for about 1 to 2 hours each bubbling. When the washed water was cleared finish the washing. The valve takes out upper layer of washed biodiesel from the bottom milky water, which contains dissolved KOH. The biodiesel was heated up to 110 °C for 10 minutes to remove excess water if present. Then biodiesel was cooled down to room temperature before use and presenting a 94 per cent yield.

8.10 Transesterification (biodiesel reaction for castor oils < 2.5%FFA)

The transesterification process can be summarized in the following steps:

1) Heat castor oil up to 140° F
2) Titrate the oil (determine how Potassium Hydroxide to add)
3) Mix the Potassium Hydroxide and methanol to make methoxide
4) Mix the methoxide with the castor oil
5) Drain Glycerin
6) Wash biodiesel

8.10.1 Esterification (pretreatment where FFA > 2.5%)

Esterification is done as a pretreatment step to the transesterification procedure when the FFA content is higher then 2.5%. In practice, it is a bit more complicated to implement then transesterification. A byproduct of the process is water, which impedes the reaction. As there is more FFA in the oil, more methanol percentage wise must be added to compensate for the water. To overcome this, industrial producers use counter current reactors that enable a continuous flow of high FFA oil in and water out.

8.10.2 Properties of Biodiesel

Generally, the properties of biodiesel and especially its viscosity and ignition properties are similar to the properties of fossil diesel. Although the energy content per liter of biodiesel is about 5 to 12 % lower than that of diesel fuel, biodiesel has several advantages. For example the cetane number and lubricating effect of biodiesel, important in avoiding wear to the engine, are significantly higher. Therefore the fuel economy of biodiesel approaches that of diesel. Additionally, the alcohol component of biodiesel contains oxygen, which helps to complete the combustion of the fuel. The effects are reduced air pollutants such as particulates, carbon monoxide, and hydrocarbons. Since biodiesel contains practically no sulfur, it can help reducing emissions of sulfur oxides[5].

Biodiesel is sensitive to cold weather and may require special anti-freezing precautions, similar to those taken with standard diesel. Therefore winter compatibility is achieved by mixing additives, allowing the use down to minus 20 °C. Another problem is that biodiesel readily oxidizes. Thus long-term storage may cause problems, but additives can enhance stability. Biodiesel also has some properties similar to liquid fuels are easy for transport, and can be handled with relative ease. Also they are relatively easy to use for all engineering applications, and home use. Biodiesel are also used most popularly in Internal Combustion engines. Some technically important properties are: flash point, gross calorific value, kinematic viscosity, density, acid value, free fatty acid content, etc.

8.11 Performance and Emission Test of Castor Methyl Ester on Single Cylinder Diesel Engine Test Rig

Using different combinations of castor methyl ester with diesel, test trials were carried out for engine performance on VCR engine test setup single cylinder, four stroke diesel engine test rig (Computerized) product code-234. Effect of different combinations i.e. 100 per cent diesel oil (B0), 05 per cent Castor methyl ester and 95 per cent diesel (B05), 10 per cent castor methyl ester and 90 per cent diesel (B10), 20 per cent castor methyl ester and 80 per cent diesel (B20) evaluated by conducting fuel consumption and power tests as per EMA standard test procedure.

The setup for the study consists of single cylinder, four stroke, variable compression ratio (VCR) diesel engine connected to eddy current type dynamometer for loading (Fig 8.3). The detailed specifications of the engine used are given in Table 8.5. Windows based Engine Performance Analysis Software Package "Enginesoft" was taken for on line performance evaluation. The NOx emission by the combustion of biodiesel was measured by online flue gas analyser. The tests were conducted at the rated speed of 1500 rpm at different loads. The engine was started with standard diesel fuel and warmed up. The warm up period ends when cooling water temperature is stabilized. Then fuel consumption, brake power, brake specific fuel consumption, brake specific energy consumption, brake thermal efficiency and exhaust gas temperature were measured. Same procedures were repeated for different blends of castor methyl ester.

Table 8.5: Specification of The Engine Used.

Make and model	Kirloskar diesel engine
General details	4-Stroke, water cooled, variable compression ratio engine, compression ignition
Number of cylinder	Single cylinder
Bore	87.5 mm
Stroke	110 mm
Swept volume	661 cc
Rated output	3.5 kW at 1500 rpm
Compression ration	17.5
Rated speed	1500 rpm
Temperature sensor	RTD PT 100 and K-type thermocouple
Load indicator	Digital, range 0–490.5 kN
Dynamometer	Type – eddy current, water cooled
Load sensor	Strain gauge load cell
Fuel flow transmitter	DP transmitter
Air flow transmitter	Pressure transmitter
Rotameter	Pressure transmitter

Fig. 8.3: Variable Compression Ratio (VCR) Engine Test Setup

The engine performance was analysed with different blends of biodiesel and was compared with mineral diesel. It was concluded that the lower blends of biodiesel increased the break thermal efficiency and reduced the fuel consumption. The exhaust gas temperature increased with increasing biodiesel concentration. The results proved that the use of biodiesel (produced from castor seed oil) in compression ignition engine is a viable alternative to diesel.

8.11.1 Emission Test

The emission parameters during the test included CO, CO_2, NO_X, O_2, and smoke density. The relationship between exhaust gases (CO, CO_2, NO_X, O_2, and smoke density) with change in load and time of operation were measured. The emission test was carried out at 20 min of interval at each load for different blends with the help of exhaust gas analyzer. The exhaust gas analyzer was calibrated and the readings were taken at 2 h of engine operation by inserting exhaust gas analyzers probe in exhaust outlet for each blend. In the initial inception, the engine was started with pure diesel fuel and warmed up. The warming period ends when the cooling water temperature is stabilised. Then, the exhaust gas temperature and different exhaust emissions like CO_2, CO and NOx, smoke density, O_2 and exhaust gas temperature were measured at a rated speed. A similar procedure was repeated for the CME blends B5, B10 and B20.

8.11.2 Carbon Monoxide (CO) Emissions

Fig.8.4 indicates that it is a general trend that CO emission decrease when diesel fuel is substituted with biodiesel. All the blends generally produce a low amount of CO emissions at a light load and give more emissions at higher load conditions. Biodiesel has about 11% of O_2 content in it, this helps in the complete combustion of the fuel. Hence, the CO emission level decreases with an increasing biodiesel percentage in the fuel. In the case of B0, the CO emission is higher than that of the biodiesel blends. The reduction of the CO at a maximum load (145.28 kN) in B05, B10 and B20 averaged 13.75%, 25.02% and 28.79%, respectively, compared to diesel (B0).

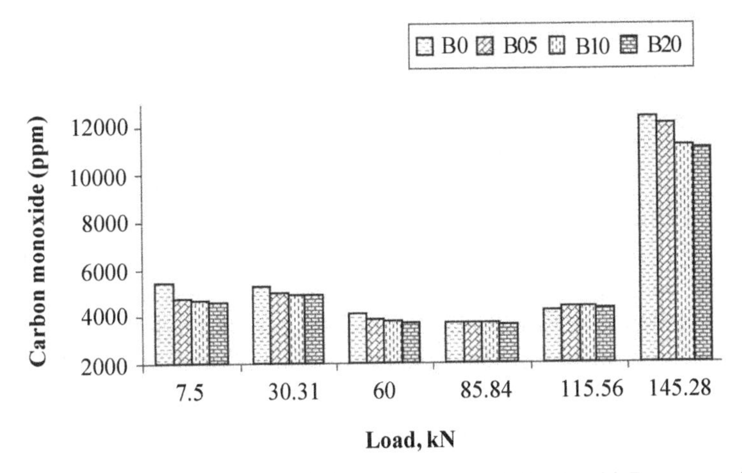

Fig. 8.4: The CO Emissions of Diesel and Biodiesel Blends with Respect to the Load

8.11.3 Carbon Dioxide (CO_2) Emissions

The Fig.8.5 compares the CO_2 emission of the various blends used in the diesel engine. The CO_2 emissions increased with an increase in the load, as expected. The biodiesel blends emit low amounts of CO_2 in comparison with B10. B20 emits a larger amount of CO_2 compared to B0. A bigger amount of CO_2 in the exhaust emissions is an indication of the complete combustion of fuel. This supports the higher value of exhaust gas temperature. The main difference in ester-based fuel and diesel is the O_2 content and cetane number. Since ester-based fuel contains a small amount of O_2 and that acts as a combustion promoter inside the cylinder, it results in better combustion for B05 than B0. The combustion of fossil fuel produces CO_2, which gets accumulated in the atmosphere and leads to many environmental problems. The combustion of biofuels also produces CO_2, but crops readily absorb these and, hence, the CO_2 level is kept in balance[6-8].

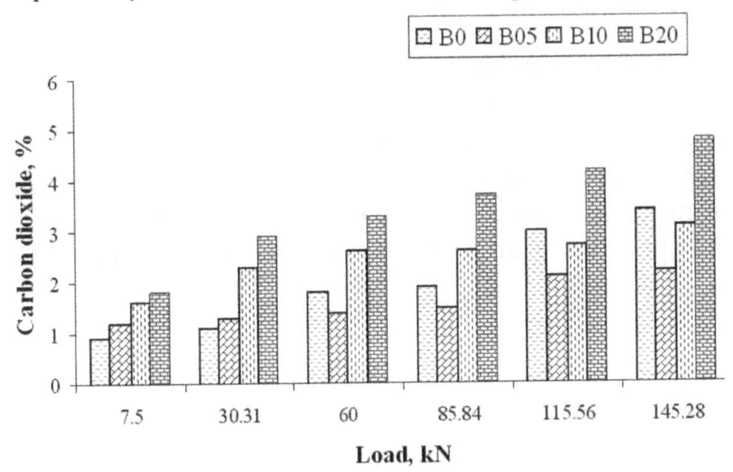

Fig. 8.5: The CO_2 Emissions of Diesel and Biodiesel Blends with respect to the Load

8.11.4 Nitrogen Oxide (NOx) Emissions

Fig. 8.6 shows the NOx emission for B0 and different biodiesel blends. It is well known that vegetable-based fuels contain a small amount of nitrogen. This contributes towards NOx production. In the case of B10, the NOx emission is lower than that of B0. The NOx concentration increased with the increase in the load and attains maximum at maximum load for all blends. In the case of B0, B05, B10 and B20, the recorded emissions were 683, 686, 638 and 694 ppm, respectively, for the 145.29 kN load. NOx emissions are dependent on engine combustion chamber temperatures, which in turn, are indicated by the prevailing exhaust gas temperature. With the increase in the value of the exhaust gas temperature, the NOx emissions have also increased. That is why biodiesel-fuelled engines have the potential to emit more NOx compared to diesel-fuelled engines[9].

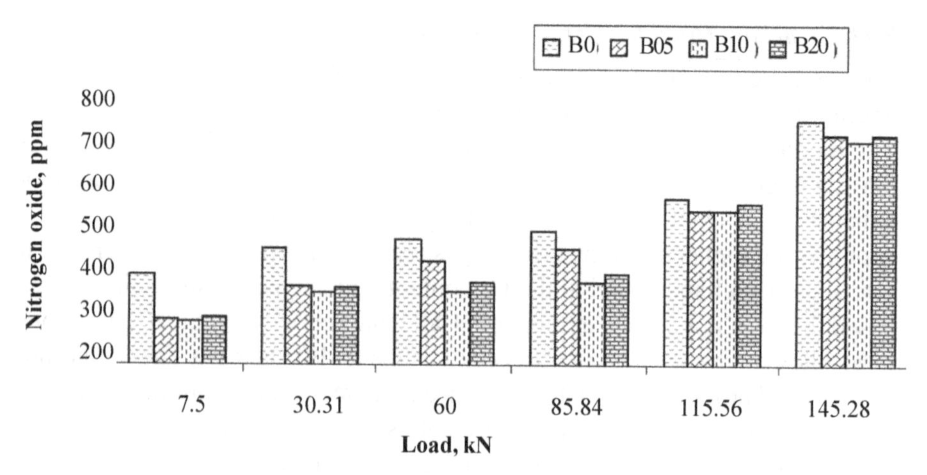

Fig. 8.6: The NOx Emission of Diesel and Biodiesel Blends with respect to the Load

8.11.5 Oxygen (O_2) Emissions

The average value of O_2 of the engine operation of each blend was plotted against different loads. Fig. 8.7 represents the relationship between the change in the load and the average O_2 emission at different loads. From the Fig. 8.7, it is observed that O_2 emissions decrease with an increase in load conditions. This drop may be attributed to O_2 having been consumed during the combustion process

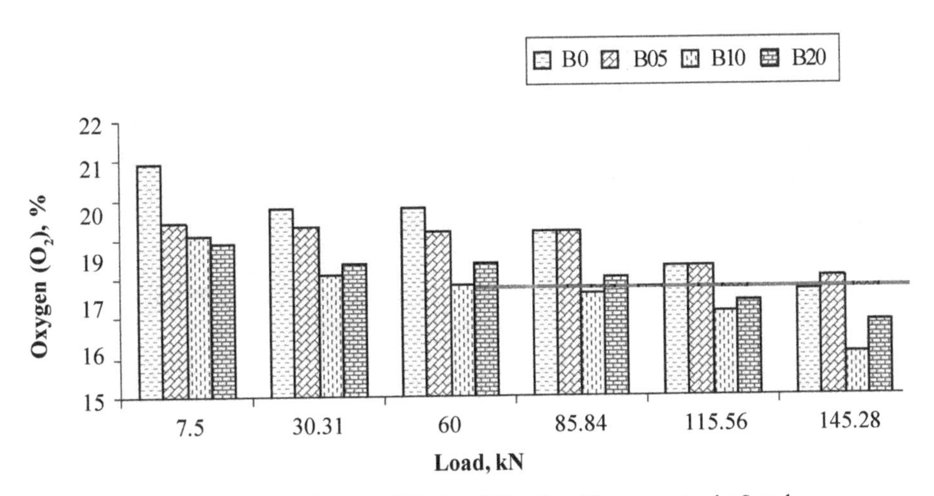

Fig. 8.7: The O_2 Emission of Diesel and Biodiesel Blends with respect to the Load

8.11.6 Smoke Density

The smoke from the engine is a function of engine load, engine performance test. Smoke density increases with an increase in the applied load. Fig. 8.8 reveals that biodiesel blends produce less smoke compared to diesel for same load conditions. During the test, it was found that B10 gave less smoke density compared B5 and B20. The smoke density of B10 at a 145.28 kN load is about 51.3% compared to 56.4% in B0.

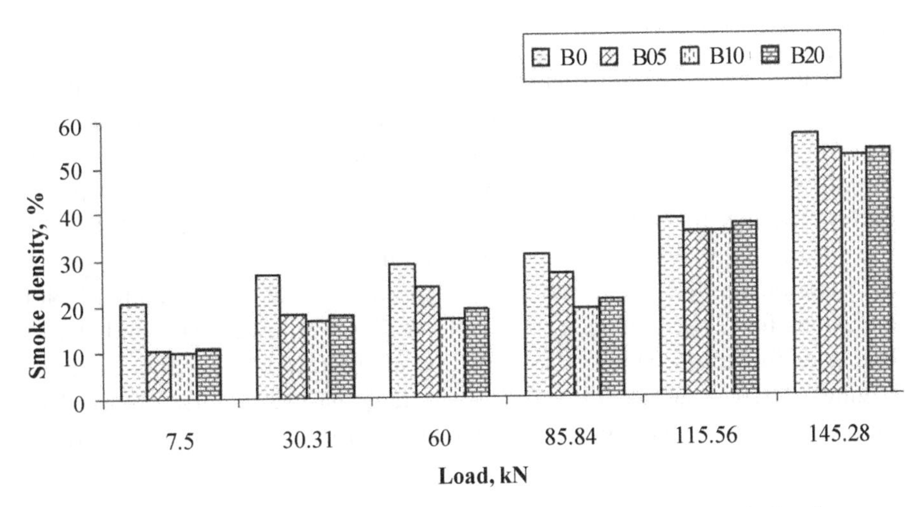

Fig. 8.8: The Smoke Density of Diesel and Biodiesel Blends with respect to the Load

8.11.7 Exhaust Gas Temperature

The variation of the exhaust gas temperature with respect to the applied loads for different blends is shown in Fig. 8.9. Up to B05, the exhaust gas temperature is lower, thereafter it increased with an increase in the blends. This reveals that effective combustion is taking place in the early stages of strokes and there is a reduction in the loss of exhaust gas energy.

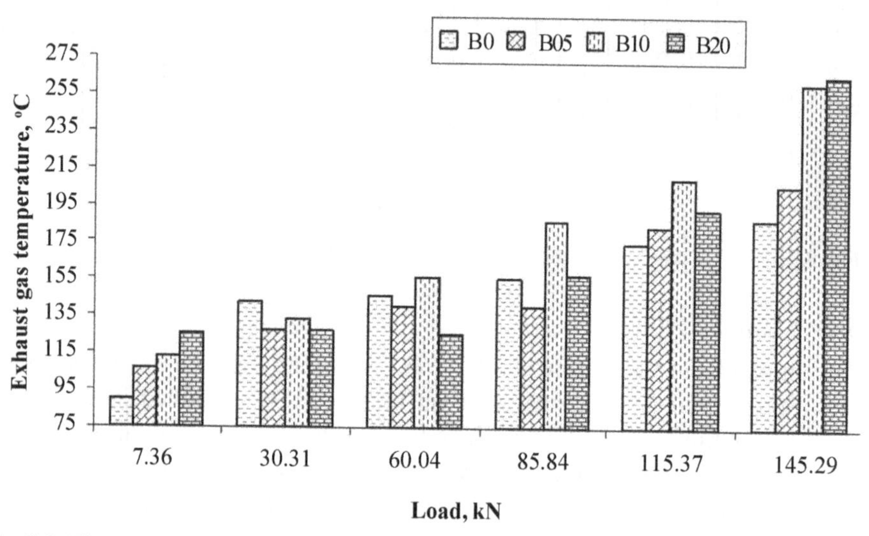

Fig. 8.9: The Exhaust Gas Temperature of Biodiesel Blends with respect to the Load

8.12 Concluding Remarks

A biomass-based fuel was obtained by the transestrification of castor oil with methanol in presence of KOH. Based on the results of this study, it was found that the important properties of CME are quite close to that of diesel. The density and calorific value of CME were found to be very close to that of diesel. The density of the CME was found to be lower than that of diesel. It was also found that CO emissions were reduced in the Compression Ignition (CI) engine fuelled with CME as compared to diesel. The trends of NOx emissions for CME blends are the same as diesel at lower loads and slightly higher at full loads. Hence, CME can be alternately used as fuel for diesel engines. There is an urgent need to encourage the use of currently available biofuels as an intermediate step to prepare the world economy for more efficient alternatives in the transport sector.

References

[1] Conceicao MM, Roberlucia A. Candeia RA, et al. Rheological Behavior of Castor Oil Biodiesel. *Energy & Fuels*.2005:19; 2185-2188.

[2] Raghuraman V, Ghosh S. A report on Indo-U.S. Cooperation in Energy – Indian Perspective. www.acus.org/docs/0303-IndoU.S._Cooperation_Energy_ Indian_Perspectives.pdf 2003 (Assess on April 2, 2008)

[3] Castor Seed Seasonal Report (2008) Karvys special reports 20080424-02. www.karvycomtrade.com Accessed 20 Mar 2009.

[4] Castor outlook (2005) Karvys special reports 20051119-01. www.karvycomtrade.com. Accessed 20 Mar 2009.

[5] BioFuel Technology Handbook-by Dominik Rutz, Rainer Janssen WIP Renewable Energies-2007)

[6] Puhan S, Vedaraman N, Sankaranarayanan G, Bharat Ram BV. Performance and emission study of Mahua oil (madhuca indica oil) ethyl ester in 4-stroke natural aspiration direct injection diesel engine. *Renewable Energy*. 2005; 30: 1269–1278.

[7] Lapuerta M, Armas O, Fernandez RJ. Effect of biodiesel fuel on diesel engine emission. Progress in Energy and Combustion Science 2008; 34: 198–223.

[8] Ramadhas AS, Muraleedharan C, Jayaraj S. Performance and emission evaluation of a diesel engine fuelled with methyl esters of rubber seed oil. *Renewable Energy*. 2005; 31; 1789–1800.

[9] Chhina R, Verma SR, Sharda, A. Exhaust emission characteristics of an un-modified diesel engine operated on bio-diesel fuels. *Journal of Agricultural Engineering*. 2005; 42: 38–43.

9

Bioethanol Production

9.1 Introduction

The rapid growth of industrialization and population significantly affect on the demand of ethanol as an alternate fuel. Generally, conventional crops like sugarcane and corn become unable to meet the demand of bioethanol, due to their primary role in food and feed. Therefore, lignocellulosic biomass from agriculture waste is attractive precursor for bioethanol production. Biomass is renewable organic material, abundant and cost effective. Ethanol production from agriculture waste has been projected an economically attractive option for industrial sector. However, sugar and starch derived ethanol have a more theoretical yield as compared to lignocelluloses, theses sources are insufficient for bioethanol production. The major agricultural wastes including rice straw, corn straw, wheat straw and sugarcane bagasse etc. can subsequently used for ethanol production. Rice straw has a highest worldwide potential for bioethanol production (205 giga liter) followed by wheat straw (104 giga liter), corn straw (58.6 giga liter), and finally for sugarcane bagasse (51.3 giga liter)[1].

A bio fuels is environmentally friendly options for power generation. Most of the fossil fuels are biological in nature. These are plant forms that, typically, remove carbon dioxide from the atmosphere and give up the same amount when burnt. The bio fuels are therefore considered to be "CO_2 neutral", not adding to the carbon dioxide level in the atmosphere. The type of bio fuel used will depend on a number of factor, chief amongst them being the available feedstock and the energy that can be used locally.

India import 70% of the oil it uses, and the country has been hard by the increasing price of oil, uncertainty and environmental hazards that are concerned with the consumption of fossil fuels. In this context, bio fuels constitute a suitable alternative source of energy for India. There are two examples of bio fuels are Ethanol and Bio diesel (Renewable Diesel)

Ethanol can be made from biomass material containing sugar, starches, or cellulose (starch and cellulose are complex from of sugar)

9.2 Composition and Properties

The alcohol molecular structure includes an (OH), or hydroxyl radical which is responsible for high solubility in water and high latent heat of vaporization. This water like characteristics are most apparent in the alcohol of low molecular weight methanol and ethanol, because the (OH) radical predominates over their short hydrocarbon chains. They are least apparent in the alcohols of high molecular weight, tertiary butyl or heavier alcohols, because their longer hydrocarbon chains predominate over the (OH) radical. These characteristics can be advantageous or disadvantageous depending on what function the alcohol it to serve. Certain properties of these alcohols are listed in Table 9.1

Table 9.1: Properties of Methanol and Ethanol[2]

Property	Methanol	Ethanol
Chemical Formula	CH_3OH	C_2H_5OH
C/H mass	3.0	4.0
Relative molecular mass	32.042	46.068
Carbon, percent by wt	37.5	52.0
Hydrogen, percent by wt	12.6	13.2
Oxygen, percent by wt	49.9	34.7
Density at 20 °C, k/m³	792.0	789.0
Boiling point °C	64.96	78.32
Freezing point °C	-94.9	-117.6
Heat of vaporization kJ/kg	1101.1	841.5
Thermal conductivity at 20 °C W/mK	0.212	0.182
Gross heating value kJ/kg	22680	29770
Stoichiometric air/fuel ratio (mass)	6.46	8.99

Example 9.1: Estimate the pure sugar required in kilogram to produce a liter of pure ethanol. The density of pure ethanol is about 789 kg/m^3

Solution

The theorical equation to produce ethanol from pure sugar is as follow:

$$C_6 H_{12} O_6 + \text{Yeast} \rightarrow 2C_2H_5OH + 2CO_2 + \text{Heat}$$

Molecular weight of pure sugar

$$C_6H_{12}O_6 = (12 \times 6) + (12 \times 1) + (16 \times 6) = 180$$

Molecular weight of pure ethanol

$$C_2H_5OH = (12 \times 2) + (5 \times 1) + (16 \times 1) + (1) = 46$$

One mole of sugar produce two moles of ethanol.

Thus, molecular weight of ethanol in equation = 94

Pure glucose required to generate one liter of ethanol:

$$=\frac{\text{Molecular weight of pure sugar}}{\text{Molecular weight of ethanol}}\times\text{Density of ethanol}$$

$$=\frac{180\text{ kg}}{92\text{ kg}}\times\frac{0789\text{ kg}}{\text{liter}}=1.54\frac{\text{kg}}{\text{liter}}$$

Hence 1.54 kg of pure sugar is needed to produce 1.0 liter of pure ethanol.

9.3 Feedstocks for Bioethanol Production

Bioethanol is usually produced from the agricultural produces which contain sugar. The agricultural produce extensively used for bioethanol production can be classified in two categories. In first category, agricultural produce containing sugar and starch, and lignicellulosic sugar containing fall in second category[3]. In the present context ethanol produce using yeast to ferment the starch and sugars in sugar beets, sugar cane, and corn. The starch contained in corn kernels is fermented into sugar, which is further fermented into alcohol.

9.4 Ethanol from Starch Materials

There are considerable amount of farm crops capable of being extensively used as raw material for fermentation process leading to ethanol production. A variety of raw material drawn from the vegetable world, and widely grown crops are capable of fermentation into ethanol. Starch based feedstock include a variety of cereals, grains, and tuber crops.

Starch content of some of the major sources are presented in Table 9.2. The starchy feed stocks can be hydrolysed to get fermentable sugar syrup and give an average yield of upto 42 liters of ethanol per kg of feedstock[4]

Table 9.2: Starch Based Feedstocks for Ethanol Production[2]

S.No.	Feedstock	Starch Percentage
1.	Corn	60 - 68
2.	Sorghum	75 - 80
3.	Rye	60 – 63
4.	Cassava	25 – 30
5.	Rice	70 – 72
6.	Barley	55 – 65
7.	Potato	10 - 25

9.5 Ethanol Production Processes

The process of producing ethanol can be schematized as shown in Fig 9.1:

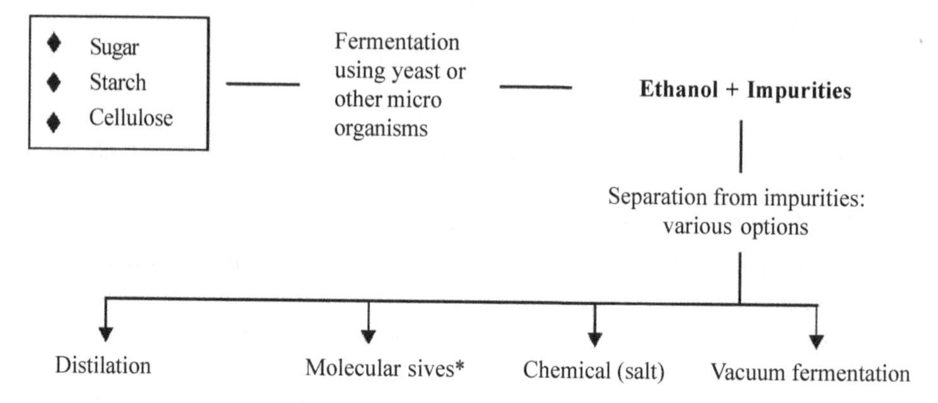

Fig. 9.1: Schematic View of Ethanol Production

* Today, almost all ethanol plants use molecular sieves for dehydration. The technology alone reduces energy use by 10 per cent per litre of ethanol produced.

Two methods are currently used to produce ethanol from grain: wet milling and dry milling.

Dry mills produce ethanol, distillers' grain and carbon dioxide (Fig. 9.2). The carbon dioxide is a co-product of the fermentation, and the distillers' dried grain with solubles (DDGS) is a non-animal based, high protein livestock feed supplement, produced from the distillation and dehydration process. If distillers' grains are not dried, they are referred to as distillers' wet grain (DWG).

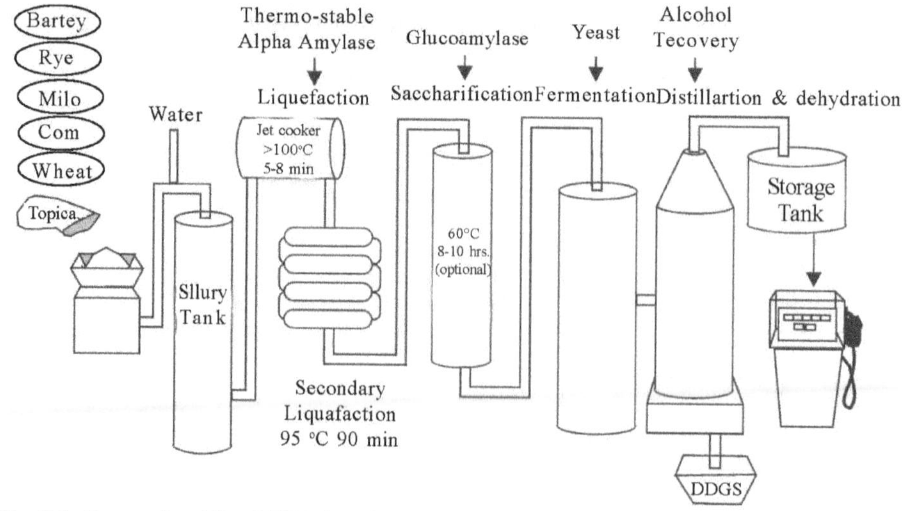

Fig. 9.2: Conventional Dry Mill Ethanol Production Process.

Wet mill facilities are 'bio-refineries' producing a host of high-values products (Fig. 9.3). Wet mill processing plants produce more valuable by-products than the dry mill process. For example, in wet mill plants, using corn as feedstock, they produce:

- Ethanol;
- Corn gluten meat (which can be used as a natural herbicide or as a high protein supplement in animal feeds);
- Corn gluten feed (also used as animal feed);
- Corn germ meal;
- Corn starch;
- Corn oil; and
- Corn syrup and high fructose corn syrups.

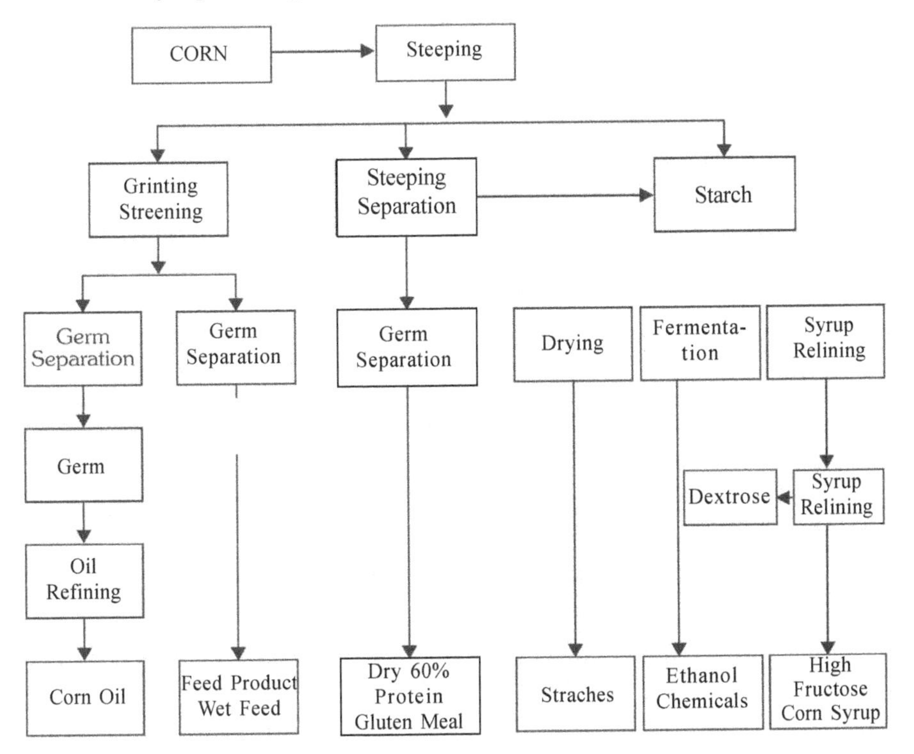

Fig. 9.3: Conventional Wet Mill Ethanol Production Process.

9.6 Pretreatment of Lignocellulosic Biomass

The main challenge in biofuel production is the pretreatment of lignocellulosic biomass. Lignocellulosic biomass is basically composed of three main constituents as cellulose, hemicelluloses and lignin. After pretreatment process the solid biomass becomes more accessible for further biological and chemical treatment. In case of raw biomass, constituents of cellulose and hemicelluloses are tightly packed by lignin layer, which restrict them for further enzymatic hydrolysis. Therefore, it is essential to have a pretreatment process to break the lignin layer to expose the constituents mainly cellulose and hemicellulose for further enzymatic action. Following are the main goals of an effective pretreatment process;

- Minimizes the crystallinity of cellulose, removes the hemicelluloses, improves the biomass surface area, break the lignin layer etc.
- Cellulose became more accessible to enzymes, therefore possible conversion of carbohydrate polymers into fermentable sugars may be achieved
- Formations of sugars by hydrolysis
- to minimize the losses and degradation of sugars
- to reduce the production cost etc.

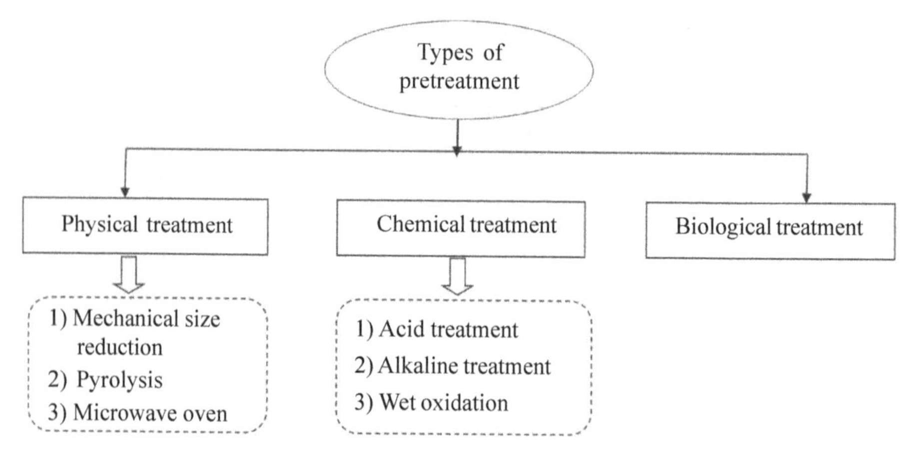

Fig. 9.4: Classification of Different Pretreatment Processes

Pretreatment process mainly classified into three categories as physical, chemical, and biological shown in Fig. 9.4

9.7 Ethanol Production from Biomass

Agro waste mainly composed of cellulose, hemicelluloses and lignin. There are main two routes for conversion of feedstock to ethanol, which can be referred as sugar route and the syngas route. The basic concept of ethanol production is shown in Fig. 9.5. In sugar conversion route, the cellulose and hemicelluloses content initially converted into fermentable sugar, which is again fermented to produce bioethanol. The fermentable sugar mainly composed of glucose, arabinose, xylose, mannose, and galactose. These sugar components were generated after hydrolysis of hemicelluloses and cellulose in the presence of either enzymes or acids. On another side, in case of syngas route, pretreated biomass is subjected to gasification process for the production of syngas or producer gas. In gasification process, biomass is heated at high temperature in no oxygen or limited supply of oxygen required for complete combustion. It subsequently produce a syngases which contains mostly hydrogen and carbon monoxide, normally gas called as synthesis gas or producer gas or syngas. These syngases are further can be fermented using some specific microorganisms or catalytically converted into ethanol. However, in case of sugar conversion route only carbohydrates are used for ethanol production, while in case of syngas platform all the lignocelluloses of the biomass are converted into ethanol[5].

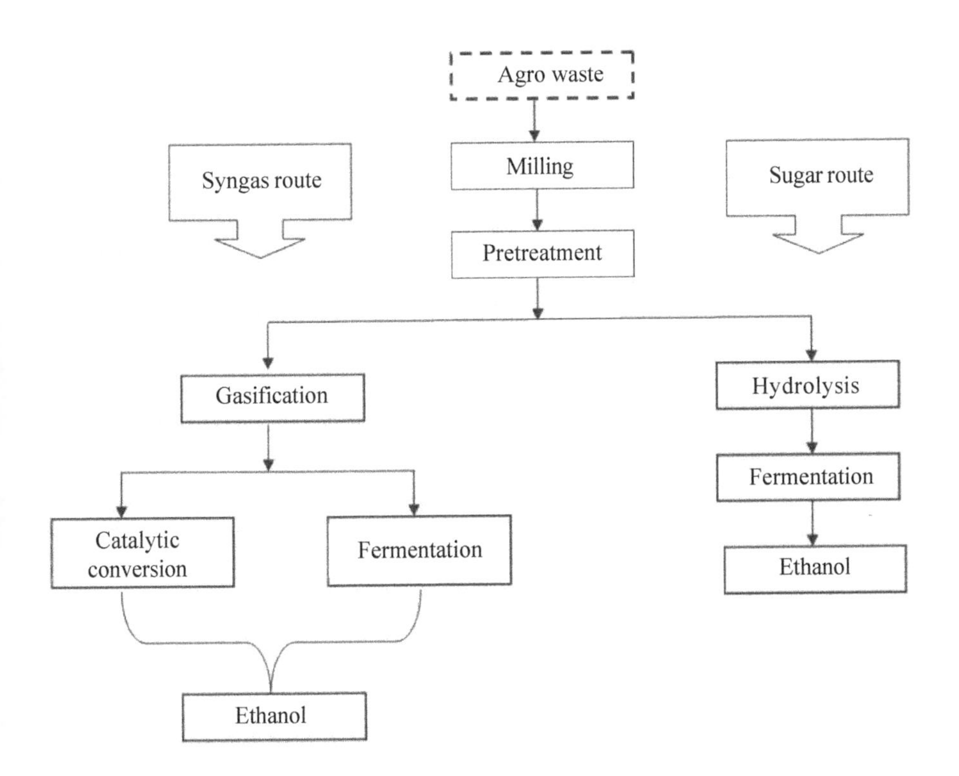

Fig. 9.5: Flow Chart of Ethanol Production from Agro Waste

9.8 Applications of Ethanol

Most probably ethanol is considered as cost effective blending fuel with a small portion of volatile fuel like gasoline. Thus, blending of bioethanol with diesel or gasoline has been used including;

- E85G (85% ethanol, 15% gasoline)
- E15D (15% ethanol, 85% diesel)
- E5G to E26G (5-26% ethanol, 74-95% gasoline)

In addition, bioethanol can be used as a;

- As an alternative to gasoline
- As a transport fuel
- For power generation after thermal combustion
- In a fuel cell
- For application in cogeneration system
- Application in chemical industry
- Reduce the greenhouse gas emission

References

[1] Sarkar N, Ghosh SK, Bannerjee S, Aikat K. Bioethanol production from agricultural wastes: an overview. Renewable energy 2012; 37(1), 19-27.

[2] Mathur HB. Alcohols – the bio-solar fuel. A state of art report submitted to DNCES, Ministry of Energy, Govt of India 1988.

[3] Vohra M, Manwar J, Manmode R, Padgilwar S, Patil S. Bioethanol production: Feedstock and current technologies. Journal of Environmental Chemical Engineering 2014; 2: 573–584.

[4] Paul JK. Large and small scale Ethyl Alcohol Manufacturing process from agricultural raw materials. Noyes Data Crop – New Jersey USA 1980.

[5] Binod P, Sindhu R, Singhania RR, Vikram S, Devi L, Nagalakshmi S, et al. Bioethanol production from rice straw: an overview. Bioresource technology 2010; 101(13): 4767-4774.

10

Densification Technology

10.1 Introduction

Biomass fuels are a potential source of renewable energy. One of the major barriers to their widespread use is that biomass has a lower energy content than traditional fossils fuels, which means that more fuel is required to get the same amount of energy. When combined — low energy content with low density - the volume of biomass handled increases enormously. Compaction or densification is one way to increase the energy density and overcome handling difficulties.

Densification means compaction of loose material or to increase density of loose biomass so that its volumetric efficiency can be increased. Densification essentially involves two parts; the compaction under pressure of loose material to reduce its volume and to agglomerate the material so that the product remains in the compressed state. The resulting solid may be a briquette, a pellet and a cube. It will be briquettes if roughly, it has a diameter greater than 30 mm. Smaller sizes are normally termed pellets though the distinction is arbitrary. The process of producing pellets is also different from the typical briquetting processes. The densified product can be developed in the cubical shapes as well.

If the material is compacted with low to moderate pressure (0.2-5 MPa), then the space between particles is reduced. Further, increasing the pressure will, at a certain stage particular to each material, collapse the cell walls of the cellulose constituent, thus approaching the physical, or dry mass and more, density of the material. The pressure required to achieve such high densities are typically 100 MPa plus. This process of compaction is entirely depending on to the pressure exerted on the material and its physical characteristics including its moisture content.

The reduction of material density is required for both the savings in transport and handling costs and improvement in combustion efficiency over the original material. Thus, densification saves cost of stage, transportation and increase calorific value per unit volume of material. The ultimate density of a briquette will depend on the nature of the original material and the machine used and its operating conditions. However, the ultimate apparent density of a briquette from nearly all materials is more or less constant, it will normally vary between 1200-1400 kg/m³ for high pressure processes. Lower densities can result from densification in press using hydraulic pistons or during the start-up period of mechanical piston presses, whilst even higher densities are sometimes achieved in pelletisation presses. The ultimate limit for most materials is in between 1450 -1500 kg/m³. The relation between compression pressure, briquetting process and the resulting density of the briquette is illustrated in Fig. 10.1

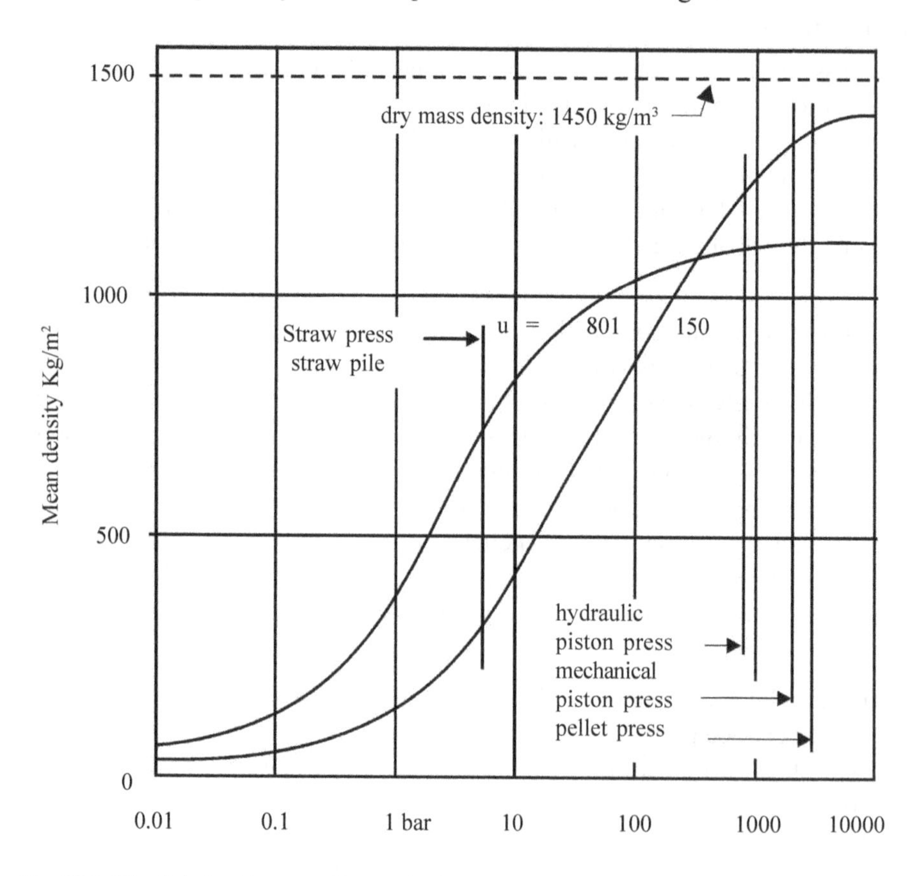

Fig. 10.1: The Relation Between Comparison Pressure and Density of Material

The apparent density of briquette are higher than its bulk or packing density as the briquettes are not generally pack perfectly. The usual reduction would be a factor of roughly two depending on the size and shape of the briquette, that is bulk densities of 600-700 kg/m³ are usual, obtained.

The bulk density of the original material like straw, which are very easy to compress even manually, is difficult to measure accurately. The lowest bulk densities are around 40 kg/m³ for loose straw and bagasse, up to the highest levels of 250 kg/m³ for some wood residues. Thus, gains in bulk densities of 2-10 times can be expected from densification process. Further, the material have to be dried after compaction in order to facilitate briquetting, thus its increases energy content per unit volume.

A binding agent is also required to prevent the compressed material for returning to its original form. This agent can either be added to the process or, when compressing ligneous material, be part of the material itself in the form of lignin. Lignin, (sulphuric lignin, is a constituent in most agricultural residues) can be defined as a thermo plastic polymer, which begins to soften at temperatures above 100°C and is flowing at higher temperatures. The softening of lignin and its subsequent cooling while the material is under pressure, is the key factor in high pressure briquetting. It is a physico-chemical process related largely to the temperature reached in the briquetting process and the amount of lignin precent in the original material. The temperature in many machines is closely related to the pressure though in some cases, external heat is applied.

There are two ways of classifying briquetting processes:

1) High, intermediate or low pressure brequetting process: this distinction is in principle, dependent on the material used but the following rough classification may be adopted:

Low pressure up to	5 MPa
Intermediate pressure	5-100 MPa
High pressure above	100 MPa

2) Whether or not an external binding agent is added to agglomerate the compressed material. Usually high pressure processes will release sufficient lignin to agglomerate the briquette, however it may not be true for all materials. Intermediate pressure machines may or may not require binders, low-pressure machines require binders for compaction.

10.2 Pre-treatment of Biomass

Prior to biomass densification, pre-treatments may be required to optimize the energy content and bulk density of the product.

Pre-treatment can include:

- Size reduction
- drying to required moisture content

- Application of a binding agent
- Steaming
- Torrefaction

10.2.1 Size Reduction

Densification process requires specific size of biomass to achieve:

- Lower energy use in the densification process
- Denser products
- A decrease in breakage of the outcome product

10.2.2 Drying

Low moisture content is desirable in biomass to improved density and durability of the fuel[1]. The desirable moisture content is in the range of 8%–20% (wet basis)[2]. Most compaction techniques require a small amount of moisture to "soften" the biomass for compaction. Above the optimum moisture level, the strength and durability of the densified biomass are decreased.

10.2.3 Addition of a Binding Agent

The compaction of biomass during densification process highly depends on inbuilt binding agents of biomass. The binding capacity increases with a higher protein and starch content[3]. Corn stalks have high binding properties, while warm-season grasses, which are low in protein and starch content, have lower binding properties[4]. Binding agents may be added to the material to increase binding properties. Commonly used binders include vegetable oil, clay, starch, cooking oil or wax.

10.2.4 Steaming

The addition of steam prior to densification can aid in the release and activation of natural binders present in the biomass.

10.2.5 Torrefaction

Torrefaction is a version of pyrolysis processes that comprise the heating of biomass in the absence of oxygen and air. Torrefaction is a pre-treatment process used to improve the properties of pellets. It can also be used as a stand-alone technique to improve the properties of biomass. Torrefaction is a mild version of slow pyrolysis in which the goal is to dry, embrittle and waterproof the biomass. This is accomplished by heating the biomass in an inert environment at temperatures of 280°C–320°C.

10.3 Technology or Devices for Densification of Biomass

10.3.1 Piston Press

In piston press, as name indicates, pressure is applied discontinuously by the action of a piston on material placed into a cylinder. They may have a mechanical coupling and fly wheel or utilise hydraulic action on the piston (Fig. 10.2). the continuous effect of piston pressure agglomerate the loose material and thus compacted to form briquettes. Briquettes produced through piston press having diameter in the range of 30 mm or greater and are formed when biomass is punched, using a piston press, into a die under high pressure.

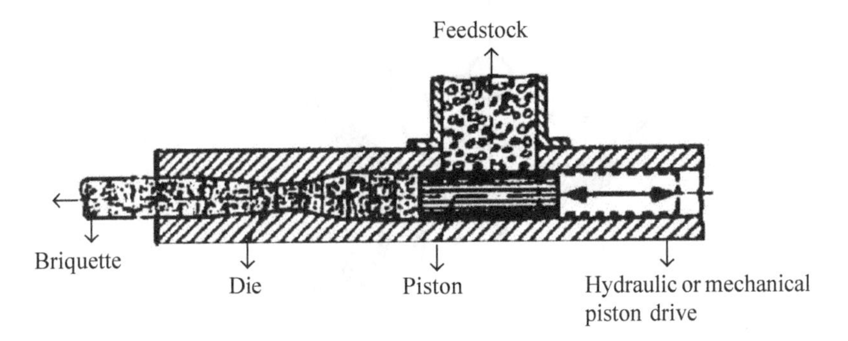

Fig. 10.2: Piston Press

10.3.2 Screw Extruder

In screw extruder, pressure is applied continuously by passing the material through a screw with diminishing volume. There are cylindrical screws with or without external heating of the die and conical screws. Units with twin screws are also available (Fig. 10.3). The volume of barrel is continuously reduces as the material is extruded outward.

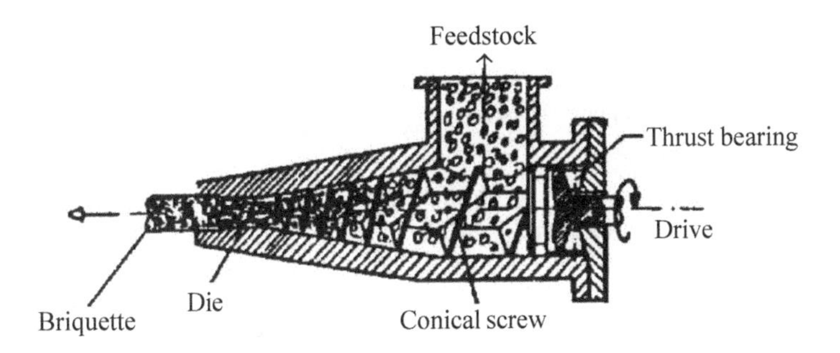

Fig. 10.3: Screw extruders

10.3.3 Pellet Press

In pellet press, rollers run over a perforated surface and the material is pushed into a hole each time when roller passes over. This continous rolling of roller over perforated surface forces loose material to form pellets. The dies are either made out of rings or disks though other configurations are possible (Fig.10.4). Pellets are easier to handle and the standard shape of a biomass pellet is a cylinder, having a length about 35 mm and a diameter around 6 to 8 mm. They are uniform in shape but can easily be broken during handling.

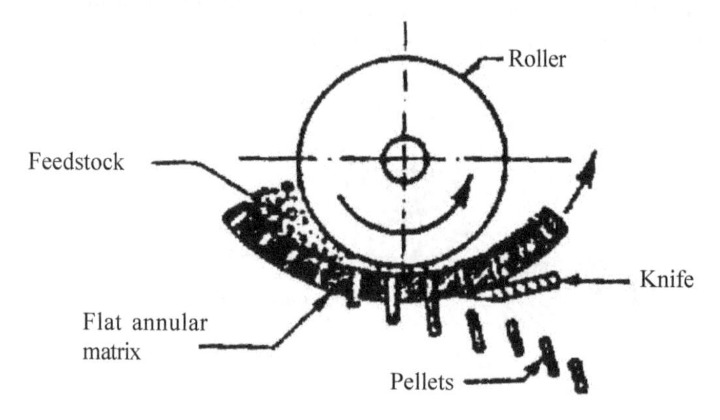

Fig. 10.4: Pellet Press

Various other types of roller-presses are also available to form briquettes, especially in making charcoal briquettes from carbonized material. A binding agent is also added in these and the process is more of agglomeration than densification as there is only a limited reduction of volume (Fig. 10.5). These roller presses are generally used for carbonized powder only. There are few other piston press available for densification process i.e mechanical screw press and pellet press etc.

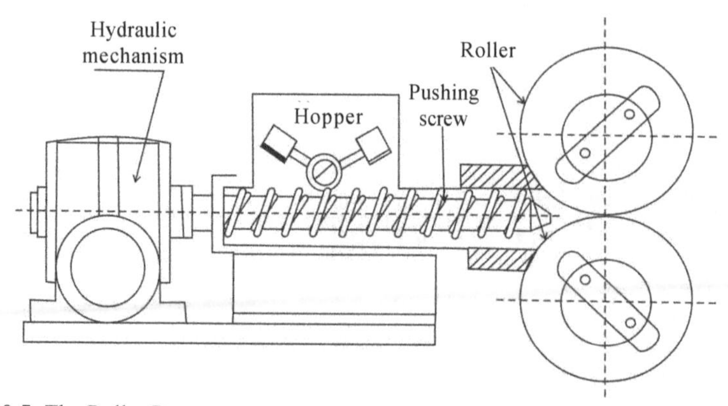

Fig. 10.5: The Roller Press

10.3.4 Mechanical Piston Press

A reciprocating piston pushes the material into a tapered die where it is compacted and adheres against the material remaining in the die from the previous stroke. A controlled expansion and cooling of the continuous briquette is allowed in a section following the actual die. The briquette leaving this section is still relatively warm and needs a further length of cooling track before it can be broken into pieces of the desired length.

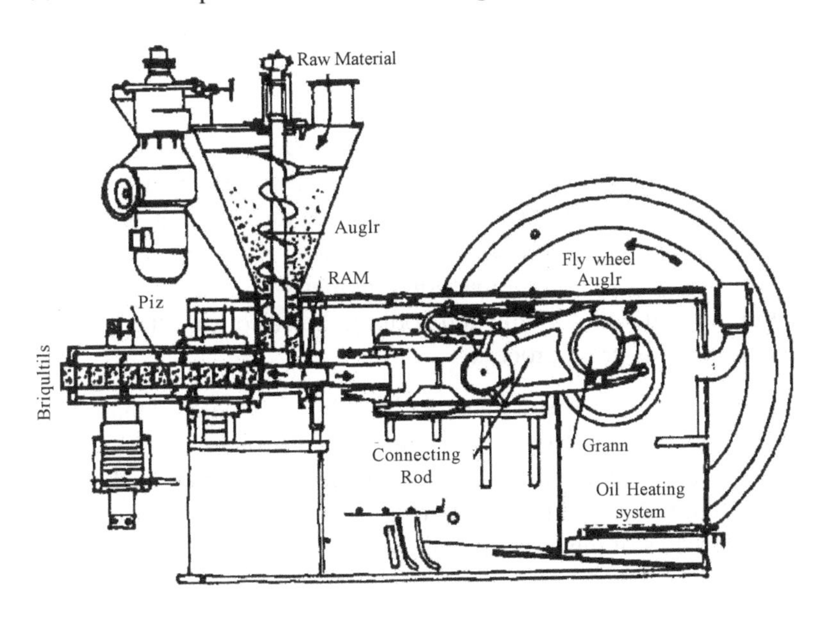

Fig. 10.6: Mechanical Piston Briquetter

The size of this type press depends on the quantity of material to be densified and nature of raw material & its heat liberation capacity during agglomeration.

In mechanical systems, the piston gets its reciprocating motion through by mounting eccentrically on a crank-shaft with a flywheel. The shaft, piston rod and the guide for the rod are held in an oil-bath. The moving parts are mounted within a very sturdy frame capable of absorbing high forces acting during the compression stroke.

The most common drive of the flywheel is an electric motor geared down through a belt coupling. A direct-drive system using an internal-combustion or steam engine are also available, which are not changing the basic design of the briquetting machine (Fig. 10.6).

The most common type of briquette press features a cylindric piston and dies with a diameter ranging from 40 -125 mm. The die tapers towards the middle and then increases again before the end. The exact form of the taper varies between machines and biomass feedstock and is a key factor in determining the functioning of the process and the resulting briquette quality.

The tapering of the dies can be in several designs. However, they may be adjusted during operation by means of narrowing a slot in the cylinder. This is achieved by either screw or hydraulic action. The optimum tapering, and thus pressure, depends on the material to be compressed.

The pressure in the compression section is in the order of 110 to 140 MPa. This pressure together with the frictional heat from the die walls, is in most cases enough to bring the material temperature up to levels where the lignin is becoming fluid and can act as a binder to produce a stable briquette. In fact, heat needs to be extracted from the process to prevent overheating. This is done by water-cooling the die. The process of compaction together with heat liberation and agglomeration coupled with binding and cooling the barrel simultaneously is overall sequence of this press process.

The capacity of a piston press is defined by the volume of material that can be fed to the piston before each stroke and the number of strokes per unit of time. Capacity by weight is dependent on the density of the material before compression. Although the nature of the original material does not alter the physical characteristics of the briquette, it does have a major impact upon the practical output of a machine.

The feed mechanism is crucial feature in the design of piston press. By means of screws or other devices, they try to pre-compress the material in order to get as efficient filling as possible. This is particularly important when using materials whose bulk density is low and which need efficient feeding to achieve more output.

The feed mechanism can, if badly mismatched with the feedstock, cause serious problems in machine operation. If undersized, voids may occur in front of the piston causing damage to the mechanism. The feeder itself may also jam if it is oversized and tries to move too much material into the piston space.

The design parameters of piston machines such as flywheel size and speed, crankshaft size and piston stroke length, are highly constrained by material and operating factors. Table 10.1 shows production capacity variations between materials.

It is likely that consumer acceptance of briquettes is also related to their size. For example, a household user cooking on a open fire would be unlikely to accept a 10 cm diameter briquette and more than a 10 cm piece of wood. Briquettes can be split or broken but this may not be accepted by the consumer and, soft briquettes may lead to crumbling. Industrial customers may, accept large whole briquettes as these conform to their usual wood sizes. This means that in designing plants to receive certain residue volumes, some attention has to be paid to the intended market in deciding, for example, on the number of machines to be used.

Table 10.1 : Production Capacity Variation between Materials

Raw material	Bulk density kg/m^3	Capacity index	Energy Index
Wood	150	100	100
Shavings	100-110	80	95
Groundnut shells	120-130	90	100

10.3.5 Hydraulic Piston Press

The principle of operation of hydraulic piston press is basically the same as with the mechanical piston press. The difference is that the energy to the piston is transmitted from an electric motor via a high pressure hydraulic oil system. Here the forces are balanced-out in the press-cylinder and not through the frame therefore, compact and light the machine can be made very. The material is fed in front of the press cylinder by a feeding cylinder i.e press-dog which often pre-compacts the material with several strokes before the main cylinder is pressurized. The whole operation is controlled by a programme which can be altered depending on the input material and desired product quality. The speed of the press cylinder is much slower in hydraulic press than with mechanical press thus which results in lower outputs. The briquetting pressures are considerably lower with hydraulic presses than with mechanical systems. The reason is the limitations in pressure in the hydraulic system, which is normally limited to 30 MPa. The piston head can exert a higher pressure when it is of a smaller diameter than the hydraulic cylinder, but the gearing up of pressure in commercial applications is modest. The resulting product densities are normally less than 1000 kg/m^3 and durability & shock resistance are naturally suffer compared to the mechanical press.

10.3.6 Screw Press

Screw press operate by continuously forcing material into a die with a feeder screw. Pressure is built up along the screw rather than in a single zone as in the piston machines. Three types of screw presses are found in the market. (1) Conical screw press (2) Cylindrical screw press with heated dies and (3) Ditto without externally heated dies.

In this type of press, drying takes place internally in the machine from the frictional heat developed in the process. A system of funnels allows the generated steam to escape from the material and the process can accept raw materials with moisture contents up to 35 %. The energy for the drying will have to be supplied through the mechanical power drive which means that the electric motors are oversized when compared to processes densifying dry material. The higher energy costs for drying with electricity compared to fuel or solar drying, plus the difficulties envisioned in installing the large motor drives in weak electricity grids, make such application unlikely (Fig 10.7).

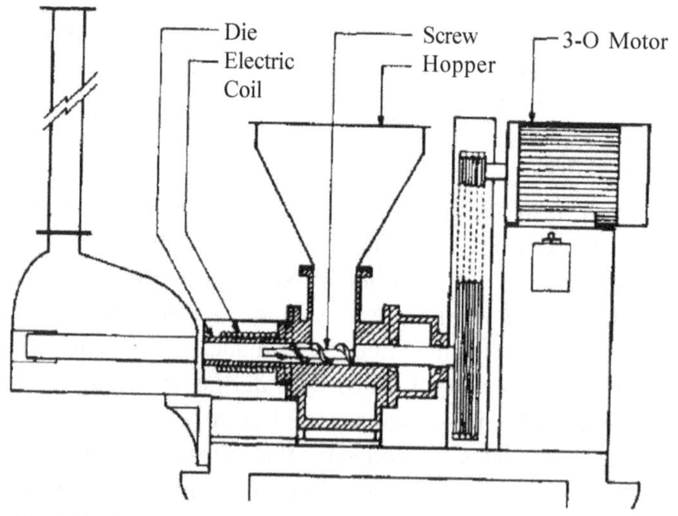

Fig. 10.7: Screw Press with Heated Die

10.3.6.1 Conical Screw Press

Conical screw presses are available in capacity of 600 to 1000 kg/m^3. It features a screw with a compression die-head. It is reinforced with hard metal inlays to resist the very high wear experienced with this type of extrudes, especially when briquetting abrasive materials. The die is either a single hole matrix with a diameter of 95 mm or a multiple 28 mm matrix. The briquetting pressure is 60 to 100 MPa and the claimed density of the product is 1 200 -1 400 kg/m^3. The machine is equipped with a 74/100 kW 2-speed motor. It is estimated that the actual average energy demand is 0.055-0.075 kW/kg/h. whereas piston press with the same output demand 0.058kW/kg/h.

The mixing and mechanical working of the material in the conical screw press is beneficial to the quality of the product. Continuous operation also aids quality as the briquettes produced do not have the natural cleavage lines as is the case with piston briquettes (Fig. 10.8).

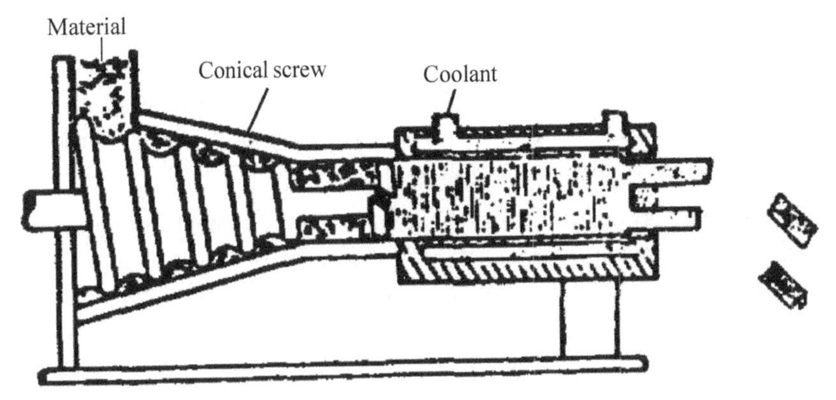

Fig. 10.8: Conical Screw Press

The main disadvantage of this type of press is the severe wear of the die head and die which results in high maintenance costs. The service life of the die head is said to be:

- With groundnut shells - 100 h
- With rice husks (estimated) - 300 h

10.3.6.2 Screw Extruders without Die Heating

This type of press is essentially a multiple hole matrix screw extruder for densifying chicken manure. It is a low-pressure process accommodating raw material at moisture content of 30%. The manure is compressed to reduce its volume by half and the product is air dried after compression for use as a boiler fuel.

10.3.6.3 Screw Extruders with Heated Dies

Essentially this type of press feeds the material from a feeding funnel, compacts it and presses it into a die of a square, hexagonal or octagonal cross-section through a screw. The briquettes have a characteristic hole through the centre from the central screw drive. The die is heated, most commonly by an electric resistance heater wired around the die. The process can be controlled by altering the temperature. The normal operating temperature is in the order of 250-300 °C. The central hole of the briquette will act as a chimney for the steam generation on account of generation of the high temperatures in the process. An exhaust is normally mounted above the exit hole from the mould where the briquettes are cut into suitable lengths. A reduction in moisture content is achieved during the formation of the briquettes. The pressure is relatively high which, combines with the high temperatures, limits the moisture content of the raw material to be used. The actual maximum moisture content depends of the raw material but is in the order of 15-20%.

Most models produce a briquette with a diameter of 55 mm and an inner hole diameter of 15 to 25 mm. Variations of the outside diameters between 40 and 75 mm can be found. A common capacity given by the manufacturers of a 55 mm machine is 180 kg/h for wood material and 150 kg/h for rice husk. Variations exist due to differences in screw design and speed.

The energy demand is consistent with these variations in capacity, ranging from 10 kW for a 75 kg/in machine to 15 kW for a 150 kg/in, both based on rice-husk briquettes. For this about 3 to 6 kW of electricity should be added which is used for heating the mould. Assuming the total of 18 kW for a 150 kg/h machine, this results in a specific energy demand of 0.12 kWh/kg.

10.3.7 Pellet Press

Pellets are the result of a process which is closely related to the briquetting processes. The main difference is that the dies have smaller diameters (usually up to approx. 30 mm) and each machine has a number of dies arranged as holes bored in a thick steel disk or ring. The material is forced into the dies by means of rollers (normally two or three) moving over the surface on which the raw material is distributed.

The pressure is built up by the compression of this layer of material as the roller moves perpendicular to the centre line of the dies. Thus the main force applied results in shear stresses in the material which often is favourable to the final quality of the pellets. The velocity of compression is also slower when compared to piston presses which means that air locked into the material is given ample time to escape and that the length of the die (i.e. the thickness of the disk or ring) can be made shorter while still allowing for sufficient retention time under pressure. The pellets are still be hot when leaving the dies, where they are cut to lengths normally about one or two times the diameter. Successful operation demands that a rather elaborate cooling or drying system is arranged after the densification process (Fig.10.9).

There are two main types of pellet press: Flat and Ring types.

10.3.7.1 Flate Pellet Press

The flat die type have a circular perforated disk on which two or more rollers rotate with speeds of about 2-3 m/s i.e each individual hole is over run by a roller several times per second. The disks have diameters ranging from about 300 mm up to 1500 mm. The rollers have corresponding widths of 75-200 mm resulting in track surfaces (the active area under the rolls) of about 500 to 7500 cm^2.

10.3.7.2 Ring Pellet Screw

The ring die press includes a rotating perforated ring on which rollers (normally two or three) press on to the inner perimeter. Inner diameters of the rings vary from about 250 mm up to 1000 mm with track surfaces from 500 to 6000 cm^2.

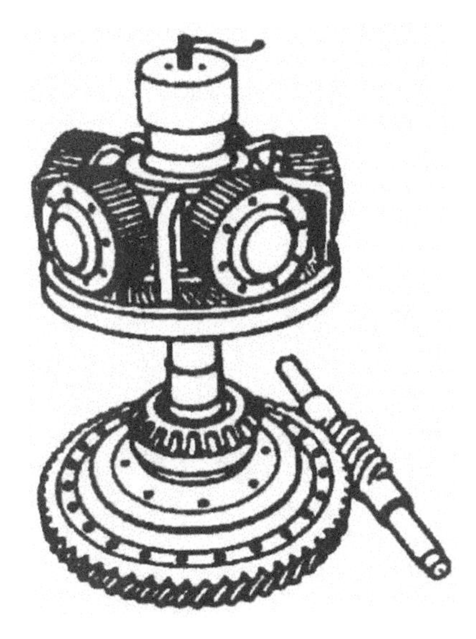

Fig. 10.9: Flat Die Type Pellet Press

10.4 Advantages of densification

The main advantages of biomass densification for combustion are:

a) Simplified mechanical handling and feeding

b) Uniform combustion in boilers

c) Reduced dust production

d) Reduced possibility of spontaneous combustion in storage

e) Simplified storage and handling infrastructure, lowering capital requirements at the combustion plant

f) Reduced cost of transportation due to increased energy density

g) Increase calorific value per unit volume of biomass

h) Additional of other higher calorific valued material to add value in original biomass

The major disadvantage to biomass densification technologies is the high cost associated with some of the densification processes.

References

[1] Shaw M, Tabil L. Compression and relaxation characteristics of selected biomass grinds. ASABE Paper No. 076183. St. Joseph, Mich.: ASABE 2007.
[2] Kaliyan N, Morey RV. Factors affecting strength and durability of densified biomass products. Biomass Bioenergy 2009; 33 (3): 337-359.
[3] Tabil LG, Sokhansani S, Tyler RT. Performance of different binders during alfalfa pelleting. Canadian Agricultural Engineering 1997; 39(1): 17-23.
[4] Kaliyan N, Morey V. Densification characteristics of corn stover and switchgrass. ASABE Paper No. 066174. St. Joseph, Mich.: ASABE 2006.